ECOLOGY OF FRESHWATER ORGANISMS AND INSECTS

Recent Advances in Environmental Ecology Series

ECOLOGY OF FRESHWATER ORGANISMS AND INSECTS

A. P. Diwan
D. K. Arora

ANMOL PUBLICATIONS PVT. LTD.
NEW DELHI - 110 002 (INDIA)

ANMOL PUBLICATIONS PVT. LTD.
4374/4B, Ansari Road, Daryaganj
New Delhi - 110 002

Ecology of Freshwater Organisms and Insects

ISBN 81-7488-122-0

PRINTED IN INDIA

Published by J.L. Kumar for Anmol Publications Pvt. Ltd., New Delhi - 110 002 and Printed at Mehra Offset Press, Delhi.

Preface

The freshwater environments include streams and lakes, as well as wetlands such as peatlands or bogs, marshes, and swamps. Aquatic features such as lakes and wetlands depend upon climatic conditions, geological events, or biological actions to develop and maintain level of water. These sites are thus geographically restricted. Estuaries develop where fresh and sea water meet and are also highly localized. Their-location is limited mainly to the mouths of rivers.

The authors have freely consulted a large number of foreign books and Journals in the preparation of the present title. Every possible effort has been made to make the book useful for student of concerned subject and as well as for general reader.

Any worthwhile criticism and suggestion for improvement would be thankfully acknowledged.

AUTHORS

Contents

Preface v

1. Freshwater Environments 1
2. Freshwater Environments in Polar Regions 42
3. Biotic Factors of Freshwater Environment 66
4. Trophic Relationship in Protozoa 117
5. Ecological Adaptations in Protozoans 153
6. Ecology of Bryozoans 178
7. Growth Forms as Adaptive Strategies 194
8. Ecology of Insects 214
9. Aquatic Insects of Soil and Water 227
10. Inter-Relationship 249
11. Insects and Man 270
12. The Fungus Gardens of Insects 332
13. Water Resources 344
14. Contamination of Water 268

1

Freshwater Environments

The aquatic systems of the terrestrial environment are, per area, among the most productive systems of the Earth. The purpose of this chapter is to examine the features and processes that contribute to their high productivity.

The freshwater environments include streams and lakes, as well as wetlands such as peatlands or bogs, marshes, and swamps. Aquatic features such as lakes and wetlands depend upon climatic conditions, geological events, or biological actions to develop and maintain bodies of water. These sites are thus geographically restricted. Estuaries develop where fresh and sea water meet and are also highly localized: their location is limited mainly to the mouths of rivers. Streams are far more common, and because they are the most ubiquitous terrestrial aquatic feature on the Earth, they will be the main focus of the chapter.

The Water Cycle

The water cycle, or hydrologic cycle, refers to the exchange of water between the ocean and atmosphere. On the Earth, the oceans, which cover 71 per cent of the surface, are the major reservoir of the water. The oceans lose water by evaporation

and gain water by precipitation. This is where the process becomes intriguing. The oceans only gain about 90 per cent of the water that the evaporated. The remainder is transferred to the land and returns to the oceans as runoff. The total amount of water involved in the cycle, about 0.04 per cent of the Earth water, represents all of the water that appears in the atmosphere (some of which appears as rain), the rivers, lakes, wetlands, soil moisture, and groundwater. At any one time, under present climatic conditions, 97.5 per cent of the water on the planet occurs in the oceans. The remaining 2.5 per cent is divided among the following categories. About 80 per cent is in the ice caps and glaciers and about 20 per cent is groundwater. Lakes contain about 0.7 per cent of the water soil moisture holds about 0.44 per cent of the water, and the atmosphere and streams contain about 0.05 and 0.005 per cent respectively of the non-ocean water. Table 1.1 lists the approxi-
it illustrates the time, percentage, and amount of each reservoir;
mate residence relatively small portion of total water involved in the land part of the cycle. The land area occupied by open

TABLE 1.1

The Major Water Reservoirs of the Earth

Reservoir	*Per cent*	*Residence time*	*Amount (1,000's km³)*
Total water	100.0		1,382,000
Oceans	97.5	1,000's years	1,347,000
Ice	2.0	1,000's—1,000,000's years	276,000
Groundwater	0.5	10—100's years	6,908
Lakes	0.0175	variable	241
Soil moisture	0.011	months	152
Rivers	0.0001	days	1.2
Atmosphere	0.001	10's of days	12.3

water also provides perspective of the water cycle. Lakes occupy about 2.5 million km² 1.7 per cent of the surface area of the continents of the Earth (Czaya 1981). They contain about 241,000 km³ of water. The major lakes of the Earth, *i.e.*, Lake Baikal of Asia the Great Lakes of North America, and the Rift Valley lakes of Africa contain almost three-quarters of the freshwater. All of the remaining lakes, ponds, and wetlandsshare the remaining portion. Streams on the other hand occupy less than 0.1 per cent of the area of the land (Hynes 1970). Annually, rivers, which occupy about one-seventeenth of the land area of lakes, discharge about 1,000 km³ of water back to the ocean, or nearly one-tenth as much water as stored in lakes. Annually, streams, with only a fraction of the depth of lakes, handle about 2.6 times more volume of water per occupied area in lakes. That in essence describes the stream—a large volume of water flowing among a narrow ribbon of band.

Streams

At any one time, streams and rivers contain about 1,230 km³, or about 3.3 per cent of the annual volume of water whch returns to the seas. Stated differently, during the course of a year, the Earth's streams transport to the sea about 30 times more water than they contain at any one time. There are many possible transfers of water from streams. Some water undergoes evaporation and some steeps as groundwater into the spaces between particles of sediment. Some of the groundwater enters the root some to become available to plants and thence undergoes evapotranspiration. The relationships between soils, plants, and the hydrologic cycle are described in greater detail in the next section. However, before these relationships are examined, the characteristics of streams will be considered.

Streams are located at virtually all latitudes and in all climates. Climate delivers the crucial ingredients which determine the regional characteristics of streams. Streams form under conditions where precipitation exceeds evaporation and infiltration. There are several types of streams. Perennial or effluent streams occur in humid climates where the supplyof water is sufficient to flow in every season. The smallest drainage basins of the upper reaches of a large effluent river system often have intermittent streams. These small drainage basins characteristically have large variations in stream flow because they have too little storage of water to continuously supply the stream with water. They may flow, for example, for only a few days after each input of precipitation. Ephemeral streams occur in arid climates where infrequent precipitation and high evaporation leads to inadequate water for sustaining stream flow over long periods of time or along extended channel courses. The water in these streams tends to soak into the channel bed and such streams are called influent streams. In reality, these categories are complicated by the fact that the Earth's largest drainage basins extend across several major climatic zones.

The Drainage Basin

Stream channels occur in a seasonably well ordered pattern. Most streams are part of a drainage network that, connected, remove water from a drainage basin. Generally, geological conditions determine the geometry of the drainage network. Thus, the drainage network provides important clues to the geologic nature of the terrain. Several common drainage patterns are shown in Figure 1.1.

For biogeographical purposes, the most important pattern of the drainage basin shows up in the longitudinal profile of a stream. Most of the major streams and rivers of the Earth have a slope profile that is steepest at its headwaters, and has an exponential decrease in slope toward the mouth.

The drainage basin of a large river consists of many smaller drainage basins containing smaller streams. A scheme has been devised to classify drainage basins according to their position within the hierarchy of the drainage network. The classes of streams are called orders. A first-order stream basin has no tributaries. It is the top of the stream profile. Its slopes tend to be the steepest and with small tributary area, it tends to

FIG. 1.1: Common drainage patterns of the Earth.

have low flows. A second-order stream basin is formed when two first-order stream basins join; a third-order stream basin occurs where two second-order basins join, and so on. Higher order streams are associated with low slopes. Because of the number of streams that have joined to create high order streams, these streams usually have large volumes of water.

The relationship between slople and stream flow rate provides an interesting insight into biologically significant properties of streams. Intuitively, the relationship may seem straightforward. That is, one might expect that on steep slopes, streams flow rapidly and on low slopes they flow slowly, and that the different sized streambed sediment at both segments matches expectations (*i.e.*, larger sized materials in steeper segments and finer materials in flat segments). Streams, however, behave very differently than intuition suggests and there are many physical processes which need to be taken into account to describe the difference between upper and lower reaches of a stream. Upper reaches of a stream are found in highlands. The water source of streams is directly related to climate. Certainly, orographic precipitation in upland areas is important in contributing water to streams, and depending on elevation or latitude, may be stored in the form of ice and snow and appear in the stream as melt water. The bed material of the stream generally includes an assortment of boulders and other coarse materials. The valley which the stream traverses is generally a narrow, steep-sided, V-shaped valley. Under these conditions, a stream encounters much resistance as it flows over the streambled. Much turbulence is apparent in the formation of various eddies and standing waves. While the water may appear to be moving swiftly, in reality, streamflow is restricted by channel resistance in its downstream progress. In the upper reaches, boulders and pebbles undergo grinding and wear as they are moved down stream (Adams 1979).

The lower reaches of a stream. *i.e.*, high-order stream, occur near base level, either at a lake or at sea level. Here, there may be expanses of flat land on either side of the stream which periodically become flooded, hence the name floodplain. The stream in these places is both wide and deep and the streambed particles consist of silts and clays. The streambed surface is thus relatively smooth, providing little resistance to streamflow which is less turbulent in appearance. In addition, the volume of the stream water in the lower reaches creates the condition whereby most of the mass of water comes into contact with only water. Only a small fraction of the volume comes into contact with the stream bottom and sides. Under these conditions, the stream velocity on low slopes is higher than it is on steep slopes in the upper reaches (Leopold *et al.* 1964). Based on a continuum of bedload characteristics, slope, amount of turbulence, and flow, the stream is naturally divided into upper and lower segments. These physical divisions have distinctive properties that are important to organisms.

Regional variation of Drainage Basins

The previous section dealt with stream characteristics that generally apply to all drainage basins. There are some characteristics that distinguish drainage basins from each other. In general, most of the major distinctions occur as a result of variations in climate and geology.

Climate clearly influences the characteristics of a given drainage basin. In both the Blue and White Nile headwaters, there is ample precipitation. In the headwaters, the Nile is a normal effluent stream which receives a portion of its flow from soil and groundwater. However, as the Nile flows across the wet-dry Tropics and into the desert the situation changes. The last major tributary entering the Nile is the Atbara, over 2000 km from the mouth of the Nile. From thereon, the Nile becomes an influent stream, whereby it feeds groundwater reservoirs and undergoes evaporation during its entire journey across the Sahara Desert. In this manner, runoff is continuously diminished from the flow that originates in the upper reaches of the stream. A stream which passes through a desert for a major portion of its length is called an exotic stream. Historically, much confusion arose about the nature of streams because of the fact that the Nile appeared to come from a desert, and even more amazingly, annually flooded the land even though it came from the desert.

The major streams of the Earth listed in Table 1.2 are not exotic. They receive ample rainfall at their source and throughout most of their course, thereby gaining volume on their route to the sea. Most of the great rivers exist in humid environments. Some of the most voluminous rivers such as the Yenisei and Lena are not located in particularly humid environments. They drain an exceptionally large region, however, and they exist in a cool climate with a relatively low rate of evapotranspiration.

These drain into interior basins such as the Dead Sea, Death Valley, Caspian Sea, and Lake Eyre. With the exception of Australia's Lake Eyre, these lakes are located below sea level. They all exist in climates where, overall, evaporation exceeds precipitation. These inland basin lakes of the Earth are the recipients of stream water from surrounding drainage basins. They illustrate the point that if a stream does not receive enough precipitation, it cannot carve a path to the ocean. Many interior basins of the Earth are intriguing because they contain a biota that is relict from a time when the drainage basin was connected to a larger network of streams during more humid times. These basins provide an opportunity to study evolutionary trends that occur in isolation, in addition to those presented by oceanic islands (Naiman & Soltz 1980).

In addition to influencing the flow volume of a stream, climate plays a related role in establishing the physical conditions that cause weathering and consequently the accumulation of various minerals in soils. Tropical soils tend to be rich in aluminium and iron oxides and poor in cations such as calcium, magnesium, sodium, and potassium. Tropical streams tend to be devoid of base minerals and are thus very soft water. During flooding the sediments carry the aluminium and iron oxides into tropical rivers. Likewise, middle latitude dry soils are enriched in a variety of base elements. The streams of these regions tend to be rich in base elements and thus are hard water. During flooding, the stream sediments are enriched in base-rich elements. In humid middle latitude climates, the soils may be heavily leached and the waters arising from them poor in base elements and therefore soft. High-latitude soils are rich in silica and the streams reflect this fact.

Streams may tend to mirror soil-forming processes but generalizations do not reflect the complexity of the actual chemical characteristics of streams. Streams are strongly influenced by regional or local geologic conditions. The water quality of the stream is determined by minerals with which it comes into contact. For example, the Colorado River passes through a salt dome in its course to the Gulf of California. In doing so, the river picks up sodium chloride. In an arid climate, the sodium chloride becomes concentrated during evaporation, and this influences water quality and the type of organisms that can exist in the river, especially during droughty years. Some rivers flow through limestone formations and pick up calcium carbonate. The dissolved carbon dioxide (CO_2) in water readily reacts with calcium carbonate to form soluble calcium bicarbonate. Calcium bicarbonate, however, stays in solution only as long as there is a certain level of free CO_2. Any condition, such as photosynthesis, which removes carbon dioxide from water also causes the precipitation of calcium carbonate. The calcium carbonate precipitate builds up a crystalline deposit around cells of algae. These deposits of calcium carbonate are called travertine (Hynes 1970). Such rivers have an unusual flora and fauna. Apart from physical conditions that arise from various regional and geological influences, drainage basins also have characteristics that result from organic activity within the basins. The next sections will further develop the connections between organisms and drainage basins.

Soil-Plant-Atmosphere Continuum

The soil-plant-atmosphere continuum refers to the many interconnections between the physical and organic components

TABLE 1.2

The Major River Basins of the Earth (Ranked by Discharge)

River	Length (km)	Basin area (km^2)	Discharge (km^3/y)	Rainfall intensity (mm/y)	Dissolved transport ($t/km^2/y$)	Solid transport ($t/km^2/y$)	Amount transported ($t \times 10^6/y$)
1. Amazon	6,303	7,049,980	3,767.8	534	46.4	79.0	290 0
2. Congo	4,373	3,690,750	1,255.9	340	11.7	13.2	47.0
3. Yangtze	5,529	1,959,375	690.8	353	NA	490.0	NA
4. Mississippi-Missouri	6,215	3,221,183	556.2	173	40.0	94.0	131.0
5. Yenisei	4,991	2,597,700	550.8	212	28.0	5.1	73.0
6. Mekong	4,186	810,670	538.3	664	75.0	435.0	59.0
7. Orinoco	2,066	906,500	538.2	594	52.0	91.0	50.0
8. Parana	3,874	3,102,820	493.3	159	20.0	40.0	56.0
9. Lena	5,853	2,424,017	475.5	196	36.0	6.3	85.0
10. Brahmaputra	1,610	934,990	475.5	509	130.0	1,370.0	75.0
11. Irrawaddy	2,152	431,000	443.3	1,029	NA	700.0	NA

12.	Ganges	1,610	488,992	439.6	899	78.0	537.0	76.0
13.	McKenzie	3,276	1,766,380	403.7	229	39.0	65.0	NA
14.	Ob	5,508	3,706,290	395.5	107	20.0	6.3	50.0
15.	Amur	4,347	1,843,044	349.9	190	10.9	13.6	20.0
16.	St. Lawrence	2,512	1,010,100	322.9	320	51.0	5.0	54.0
17.	Indus	2,898	963,480	269.1	279	65.0	500.0	68.0
18.	Zambezi	2,737	1,329,965	269.1	202	11.5	75.0	15.4
19.	Volga	3,689	1,379,952	256.6	186	57.0	19.0	77.0
20.	Niger	4,186	1,502,200	224.3	149	9.0	60.0	10.0
21.	Columbia	1,955	668,220	210.8	316	52.0	43.0	34.0
22.	Danube	2,861	816,990	197.4	242	75.0	84.0	60.0
23.	Yukon	3,186	865,060	193.8	224	44.0	103.0	34.8
23.	Fraser	1,368	219,632	112.4	512	NA	NA	NA
25.	San Francisco	3,199	652,680	107.7	165	NA	NA	NA
26.	Hwang-Ho (Yellow)	4,671	1,258,740	104.1	83	NA	2,150.0	NA
27.	Nile	6,693	2,849,000	80.7	28	5.8	37.0	10.0
28.	Nelson	2,576	1,072,260	76.2	71	27.0	NA	31.0
29.	Murray-Darling	5,427	1,072,908	12.6	12	8.2	30.0	2.3

Note: NA=not available.

of the landscape. The atmosphere provides the major energy and hydrologic inputs and also creates the conditions (wind, water vapour, and heat) for a large percentage of the output (evaporation) from a location. Plants respond to the conditions provided by the atmosphere, although the atmosphere too is influenced by albedo (the reflection of light energy from the Earth's surface) and water vapour handling properties of plants. Soils play a significant role in the hydrological cycle because of their water-holding capacities. Due to their requirement for water, various plants are both influenced by water availability and influence the amount of water in soils. Plants also play a role in soil formation. The drainage basin has been the key unit of the landscape that has provided opportunities for experimental observations of the relationships between soils, plants, and the atmosphere, and their influence on the movement of water (G. S. Brush 1982). The next two sections examine these relationships.

The Soil Factor

The basic point of reference to begin a discussion of the movement of water in the landscape is soil. Soil has varying capacity to absorb water. Infiltration of water depends upon several physical properties of soil, Particle size is the most important. If the particles are sand sized, water generally infiltrates rapidly. Smaller particles such as silt and clay tend to pack together and have much reduced infiltration rates. The existence of roots, soil fauna, and organic matter can moderate the infiltration rate considerably. Because of the relationship between volume and surface area, particle size strongly influences water-holding capacity. Larger particles such as sand adsorb water freely and then, because of a lack of surface area, be unable to retain soil moisture. Clays generally have a higher water-holding capacity because of the large surface area of the clay particles. But, if infiltration is poor, and spaces between clay particles too confined, then water-holding capacity is also reduced. Silts and loams (that is, mixtures of clay and sand-sized particles) have the largest amount of available waterholding capacity reserve for plant growth (Brady 1974) See Figure 1.2.

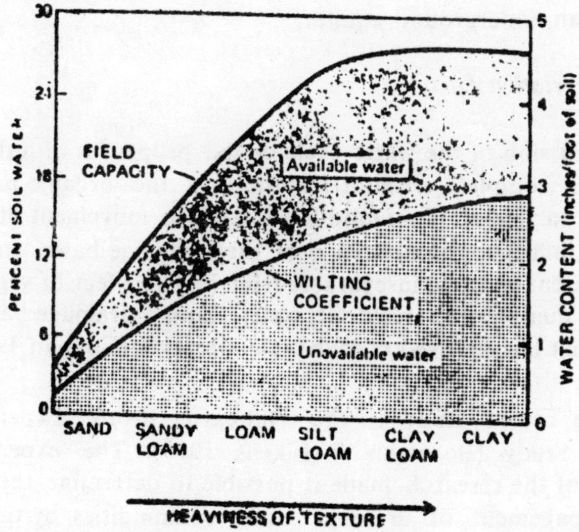

FIG. 1.2: General water-holding capacity for various soil texture classes.

Three different types of bonding between water and soil particles are importantto plants. Gravity water is that which occurs in liquid state and fills the spaces as it permeates the soil. This water is freely available to plants and moves through the soil under the influence of gravity. Capillary water clings to small particles and fills small spaces of the soil. It is also available to plants as it slowly moves into the water table. The third type of water is called hygroscopic water. It is so tightly bound to soil particles by molecular forces that it is unavailable to plants which wilt without other water. This water does not move through the soil (Brady 1974).

Soil releases water to streams and groundwater when two conditions are met. The soil must have sufficient water, and the water must be under the influence of gravity. When precipitation occurs at a rate that is equal to or greater than the rate of soil infiltration, the excess water flows under the influencce of gravity as runoff toward the stream. If more water infiltrates the soils than can be held by it, the surplus flows as groundwater and reappears in springs and streams, or enters an underground aquifer.

The Vegetation Factor

No discussion of the water cycle or the properties of a drainage basin is complete without inclusion of the organic processes and, in particular, plant modification of the movement of water. Plants influence the operation of the drainage basin to such a large extent as to be inseparable from it. The effect of vegetation on the quality and yield of water from a drainage basin is a topic that has receeived considerable attention (Colman 1953).

One recent study is the Hubbard Brook Experimental Forest Study (Bormann & Likens 1979). The experimental design of the research made it possible to determine the degree of management of drainage basin commodities by the biotic community. The commodities in this case include water and the major plant nutrients both in the soils and the atmosphere. A major purpose of the research was to measure the exchanges between bedrock, soil, plants, the atmosphere, and rivers. The research was conducted on six drainage basins in the White Mountain National Forest in New Hampshire. The drainage basins rest on impervious gneiss bedrock.

The studies began with the measurement of the water budget of the six undisturbed drainage basins, from 1955 through 1968. Through the use of specialized dams designed for measuring the flow rare of streams, runoff was measured. Also, the settling ponds behind the dams were used to measure sediment movement. Provisions were made to measure precipitation, as well as dissolved mineral load of both incoming precipitation and outgoing runoff waters. The major water budget results of Hubbard Brook Forest are as follows.

Measured inputs and outputs

Average annual precipitation =123 cm

Average annual runoff =72 cm

Difference

Average annual evapotranspiration =51 cm

From these results it can be concluded that the combination of evaporation (*i.e.*, water loss from all wetted surfaces) and transpiration (water movement from soil and groundwater to the atmosphere via plants) accounts for 41 per cent of the total annual amount of water entering the water basin (Bormann & Likens 1970). The removal of water from the soil by plants is relatively straightforward. The amount of energy required to extract soil water is small compared to the amount of energy needed to evaporate water from the leaf. Thus, up to the wilting stage, all soil water can be assumed to be equally available.

Provided there is soil moisture available to a plant, hydrostatic pressure within plant conducting tissues reaches maximum values. Stomata are openings in the leaf that control leaf ventilation. When light is not limiting the stomatal apertures of the leaf open to the maximum. In this way the leaf obtains the greatest possible flow of carbon dioxide. However, this condition also allows the rapid outflow of water vapour. Because of the limitations in the amount of carbon dioxide which a plant can obtain from the atmosphere, plants use much more water during ventilation than they actually require for physiological purposes (Penman 1970).

In the humid mid-latitude climate of New Hampshire precipitation is relatively evenly distributed throughout the year; runoff, on the other hand, is unevenly distributed. Most of the runoff occurs during March through May. Peak runoff is attributed to the spring melt of accumulated winter snow. Thus, 57 per cent of the total runoff occurs during one quarter of the year. April alone accounts for 35 per cent of the total when rain often falls on top of the snow. At the other extreme, during August, only 0.7 per cent of the annual average runoff takes place. The extremely low runoff during the mid to late summer can be explained by the effect of plant growth and transpiration on soil and groundwater (Bormann & Likens 1970). In spring, as the air and soil temperatures warm up, the vegetation begins to leaf out, ending the period of winter dormancy. Evaporative losses of water increase with rising temperatures, and plants begin to draw soil and groundwater at rates which exceed replenishment by precipitation.

During the peak temperatures of summer, soil and groundwater are being withdrawn by plants to the extent that almost no water is available for runoff. During low flow, small streams undergo a diurnal oscillation. During daylight, streamflow drops as evapotranspiration increases and during darkness streamflow increases. In autumn, as temperature decreases, large reductions occur in plant transpitation as plants shed their leaves and become dormant again. At this time of the year, rainfall replenishes the soil and groundwater reserves and runoff from the water incaeases. The ground freezes and snowfall accumulates during the winter leading to a repeat of the sequence of the annual runoff cycle.

In the Hubbard Brook study, the role of plants in modifying the movement of water was further demonstrated by deforestation of one of the six drainage basins. Following deforestation, the annual runoff of the drainage basin increased by about 30-40 per cent over the amount of runoff of the previously forested drainage basin. The greatest difference occurred during the spring and summer months when runoff values were about 400 per cent higher than expected from the basin in its vegetated state. The difference represents the amount of water which was normally transpired by vegetation in an undisturbed drainage basin. Unvegetated, drainage-basin runoff becomes more evenly distributed during the year because it closely reflects precipitation and becomes disconnected from the influence of plant processes (Likens *et al.* 1970).

Another major thrust of the Hubbard Brook study was on the biogeochemical cycle of a forested drainage basin. Comparison of disturbed and undisturbed basins demonstrates the degree of control that organisms exert on the landscape. The chemical quality of both precipitation (input) and runoff (output) were studied on the forested and disturbed drainage basins. Table 1.3 lists the precipitation inputs and runoff elements in outputs of elements in undisturbed and disturbed basins.

TABLE 1.3

Comparison of the Chemical Inputs and outputs (kg/ha) of Undisturbed and Disturbed Drainage Basins

	Forested		*Disturbed*	
Element	*Precipitation*	*runoff*	*Precipitation*	*Runoff*
Calcium	2.6	11.8	2.3	77.3
Sodium	1.5	6.9	1.3	18.1
Magnesium	0.7	2.9	0.6	16.2
Potassium	1.1	1.7	0.5	23.0
Nitrates	4.6	1.3	5.8	104.0

In the forested drainage basin, the ratios between the input and output of the elements calcium, sodium. magnesium and potassium range from 0.2 to 0.6. In the case of nitrates, the atmospheric contribution is about 3.5 times larger than the amount appearing as runoff. An important discovery of the research was that the loss of cations, *i.e.*, the nutrient ions of Table 1.3, is balanced by the input of positively charged hydrogen ions delivered by precipitation. A large proportion of the hydrogen ions are associated with sulphates and nitrates that apparently originate with industrial activities. When the water containing hydrogen ions enters the soil, the hydrogen ions replace the nutrient cations attached to the exchange sites of the soil, and the nutrients are flushed out of the soil with runoff.

Water quantity and quality variation due to disturbance can be more fully appreciated by comparing drainage-basin outputs before and after deforestation as shown in Table 1.3. Several key points can be made. Most striking is the fact that the disturbed drainage-basin biota loses control over the movement of elements in a local region. All of the nutrient elements were being released at a rate which was between 14 and 46 times greater than gained via precipitation. Many other changes were also observed in the stream system.

Physical qualities of the water undergo fundamental changes. In the forested basin, the streams are in deep shade during summer. As a result, there is a very small daily fluctuation of temperature of the stream water. Within a disturbed drainage basin, direct sunlight warms the soil during daytime and radiation losses during night cool the runoff. The overall change in warmer runoff which undergoes relatively large diurnal temperature fluctuation. It was also observed that the release of nutrients into the stream in large quantities (or eutrophication) caused severe deterioration of stream water quality. The combination of large quantities of nutrients and warm water triggered the rapid growth of algae in the stream. The algae in turn died and sank to the bottom and the bacterial decay processes consumed large quantities of oxygen. The stream became anoxic and unable to support higher forms of aquatic life. In conjuction with eutrophication there was increased erosion. In the deforested drainage basin, without a source of leaves, the layer of litter was lost from the land surface and stream bottom. These leaf-litter substances are extremely important to the stream because they form a biological barrier to erosion. Leaf-litter loss, coupled with increased runoff, caused a significant increase in stream channel erosion. In sum, the drainage basin functions as a whole system in which plants and their litter play a critical role in tying the parts of the system together (Likens *et al.* 1970).

Experimental work in several locations has revealed that not only is the drainage basin influenced by plants but that the amount of streamflow is a function of the species of plant present in the basin. One long-term, study, has been conducted in the southern Appalachian Mountains of southwestern West Virginia (Swank & Douglass 1974). The naturally occurring deciduous oak-hickory forest was cut and replaced with a stand of white pine (Pinus strobus). The study demonstrated the physical differences between basins vegetated with different species of plants. The soils within a basin vegetated with conifers tended to be much drier than those vegetated with hardwoods. Conifers photosynthesize and evaportanspire much earlier in the spring than do broadleaf deciduous trees. One measure of leaf surface is the leaf area index, which is a measure of the area of leaf surface compared wlth the area of ground the plant is occupying During the dormant season, conifers have a much larger leaf area index than broadleaf deciduous trees (9.9 compared to less than 1) and thus intercept and subsequently evaporate much more rainfall. Overall, the evaporation and transpiration by conifers results in a streamflow reduction of 20 per cent compared to drainage basins composed of the broadleaf deciduous forest (Swank & Douglass 1974). The reduced soil moisture resulting from higher transpiration rates among conifers also limits the growth of ground cover. The reduction in ground cover resulted in changes which influenced flooding cycles and water quality. During heavy precipitation, runoff of the sparsely vegetated surfaces caused an increase in soil erosion. Stream channels were deepened and slopes became steeper as an adjustment to the influence of a conifer forest. Also, the natural acidity of pine needles resulted in the leaching of cations and hence, water contained greater quantities of various elements. Type of vegetation controls basic characteristics of soil moisture. ground cover, and chemical regime, which in turn influence drainage basin characteristics such as water quality, sedimentation transport, channel depth, and angle of slope.

Vegetation cover can influence local water availability by acting as a cloud droplet inpaction network which deposits water and aerosols. Examples of this phenomenon are reported from several elevations of the northern Appalachians (Lovertt *et al.* 1982) aud Lomas de Lachay in coastal Peru (Holdridge 1967). In the subalpine forests of the northern Appalachans, inputs from cloud droplets account for about 46 per cent of the water, and from 150-430 per cent more chemical inputs than bulk precipitation. In the Peruvian forest, precipitation recorded under the tree-crown yielded 17 per cent more water than was recorded in the open. The mechanism that causes this increase in precipitation is the condensation of moisture or deposition of cloud droplets on the aerial portion of the tree. The leaves act as large nets that capture droplets of moisture which do not otherwise have a possibility of coalescence to form precipitation. Thus process attests to the capacity of the tree canopy to influence and enhance its own microclimate. When all aspects of the relationship between plants, albedo, precipitation, and soil water are taken into consideration it becomes apparent that vegetation plays a major role in modifying the Earth's climate (Shukla & Mintz 1982, Lockword 1983).

Soil Moisture Balance

The above discussion of the relationships between precipitation, vegetative cover, soil moisture, evaporation, transpiration, and runoff provides reason to consider alternative methods of classifying climate beyond the simple use of temperature and precipitation. C.W. Thornthwaite (1948) developed a method for classifying climates on the basis of energy and moisture criteria. Thornthwaite's soil water balance analysis uses month-by-month values of precipitation, potential evaporation, actual evaporation, and soil moisture surpluses or deficiencies to determine the water available for plant transpiration and growth at a given locality.

Potential evaporation (PE) is the amount of soil moisture which could be used by the vegetation if there was no limit to the supply of water. The principal determinant of PE is solar radiation. PE is usually expressed as the depth of liquid water that could be transpired per month. The value changes with

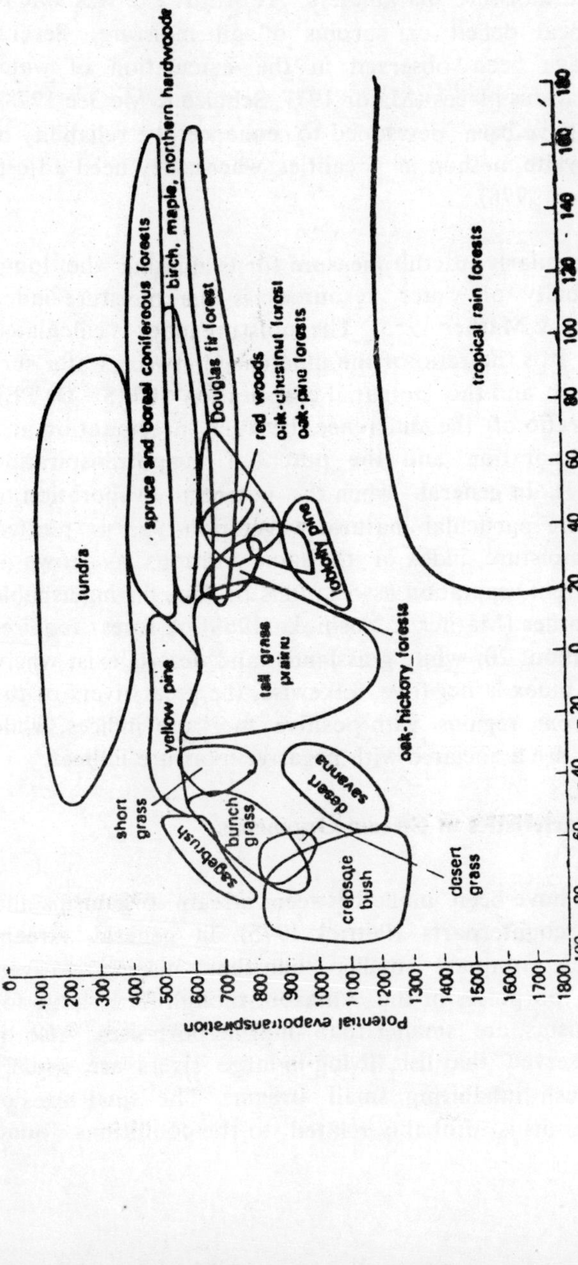

FIG. 1.3: Natural communities as a function of potential evapotranspiration and moisture index.

seasons. When soil moisture is depleted and only hygroscopic water remains, the estimated actual rate of evaporation (AE) will fall below the potential rate of evaporation. By using these two estimated moisture parameters, Thornthwaite was able to chart the local deficit or surplus of soil moisture. Several problems have been observed in the calculation of water balances in various places (Major 1977, Schulze & McGee 1978). Calibrations have been developed to enhance the reliability of the Thornthwaite method in localities where they need adjustment (Schelton 1978).

One particularly useful measure for evaluating the long-term availability of water resources is the moisture index (Thornthwaite & Mather 1955). The moisture index is calculated in two ways. It is the ratio of the difference between water surplus and deficit and the potential evaporation $100[(S-D)/PE]$; and it is the ratio of the difference between precipitation and potential evaporation and the potential evapotranspiration $100[(P/PE)-1]$. In general, when the potential evaporation of stations bearing particular natural vegetation-types is plotted against the moisture index of the same stations, as shown in Fig. 1.3, different vegetation associations become distinguishable from one another (Mather & Yoshioka 1968). A forest requires an index of about 20, while grasslands and deserts exist where the moisture index is negative. Likewise, the great rivers of the Earth arise from regions with positive moisture indices, while inland basins are associated with negative moisture indices.

General Characteristics of Stream Organisms

Comparisons have been made between stream organisms and their marine counterparts (Patrick 1975). In general, stream organisms as a group are smaller than their marine relatives. For example, porpoises of the Amazon, though fairly large for stream organisms, are smaller than marine porpoises. Also, it has been observed that fish living in large rivers are usually larger than fish inhabiting small streams. The small size of stream organisms is probably related to the conditions found in the small streams. Narrow channels and seasonal fluctuations promote the survival of small as opposed to large animals. Small streams undergo greater fluctuations in water due to small catchment area and lack of storage capacity. Large numbers of small animals have a greater likelihood of surviving the inevitable fluctuation of small streams. Under these unstable conditions small numbers of large freshwater animals would be comparatively prone to extinction.

In one way, freshwater environments of the Earth are remarkably similar. They support the same four types of insects: water striders, whirligig beetles, backswimmers, and springtails (Milne & Milne 1978). Water striders glide or dart across the water surface by rowing with their middle pair of legs. Gerris is a genus found on freshwater surfaces in all parts of the Earth. Whirligig beetles (Gyrinues and Dineules) live half in the water and half out, set up ripples as they move, and can dive or fly, and crawl on solid surfaces. Backswimmers (Nolonecta) live in ponds and the relatively quiet pools of streams. They swim belly-up and live on the underside of the surface film of the water. Their major food is the midge or mosquito. The fourth cosmopolitan surface-film aquatic insect is the springtail (Podura and Sminthurus). These latter are the most ancient surface-water insects and first appeared about 375 Ma ago. They alone inhabited the surface film until 250 Ma ago. Springtails feed on plant spores that settle onto the water surface.

Stream organisms generally have short life cycles compared with deep-lake or marine organisms. Many species have life cycles that last only one day. Since life cycles are short, the reproductive rate tends to be very high.

Stream communities tend to have long food chains that include four or five transfers of energy. Algae are important both as a food source and as a substrate for the growth of other epiphytic food sources. Rooted plants are not as important as a food source as they are as a substrate or shelter for other organisms. They thus serve to increase habitat diversity (Patrick 1975).

Streams commonly contain very large numbers of species. Many species have developed specialized organs or appendages for surviving in currents (Vogel 1981). Hard-water streams tend to support fewer species of algae, larger number of protozoa and insects, and slightly fewer species of fish than soft-water streams (Patrick 1975). Along a single river, the number of species in common during a single collection time between different collection sites tends to be higher than the number of species in common found at a single site during different collection times (Patrick 1975). Overall, there is a high species turnover in time and space. Examination of the physical characteristics that influence biotic qualities of streams provides insight into the biogeographic patterns of the stream habitat.

Photosynthesis and Respiration Rotio of Streams

The traditional way of considering a community of organisms is to regard them as a locally self-sustaining entity to which radiant energy is supplied. The energy passes through the various trophic levels, that is, the primary producer, primary consumer, secondary consumer levels, and so on, and the decay chain breaks down the organic materials so that the constituent elements can be subsequently returned to the organisms to be recycled. This biogeochemical model may be a reasonable conceptual framework for many terrestrial and even lake communities, but it is inappropriate for the stream system. In the stream, everything released into solution moves down stream and cannot be locally recycled (Hynes 1970). A generalized approach that succeeds in conveying the dynamic features of the stream is based on the longitudinal profile of the stream. There are physical features of the stream which change from the headwaters to the mouth that influence the ratio of photosynthesis and respiration. Some stream segments have almost no photosynthesis and thus a ratio of photosynthesis and respiration (P/R ratio) less than 1. The organic matter, usually leaves or needles, which supports the growth of organisms here must be imported into the stream segment. These segments are principally the domain of heterotrophic organisms. Other segments of the stream may have more photosynthesis than respiration and thus a P/R ratio greater than 1, and large numbers of autotrophs (Cummins 1975). The physical and biological properties of the upper and lower segments of a large stream are sufficiently different that the two parts of the stream have been subdivided into to rhythron and potamon.

Rhythron

The turbulent flowing stream in the elevated area, *i.e.*, headwaters and mountain streams, of the drainage basin is known as the rhythron (also called rhithron). literally flow plus elevation. Fig. 1.4 illustrates the general changes occurring in the longitudinal profile of a stream and their influence on stream organisms. The rhythron is the coldest part of the stream from the source to the point where the monthly mean temperature rises to about 68—75°F (20—24°C), where water flow is turbulent, oxygen concentration is always high, and the streambed is composed of rocks, stones, or gravel. The headwater region of a stream is typically densely vegetated and, because of the narrowness of the stream, dimly lit. The food chain of the headwater consists of inputs of organic matter from the vegetation via leaf fall. Because the stream valley is typically steep sided in the headwaters, the quantity of leaves per cubic meter of stream per year far exceeds the quantities in lower reaches of the stream. The nutrient values vary with the type of vegetation. Species of Alnus (alder) have root nodules and a symbiotic relationship with bacteria similar to legumes. Thus Alnus leaves contain about four times more nitrogen than other deciduous species. Conifers are relatively poor nutrient suppliers to streams (Hynes 1970). The rhythron supports large numbers of benthic (bottom-dwelling) organisms specialized in shredding organic matter and another group specialized in capturing course particulate organic matter (CPOM). The CPOM in turn is further digested by these organisms and

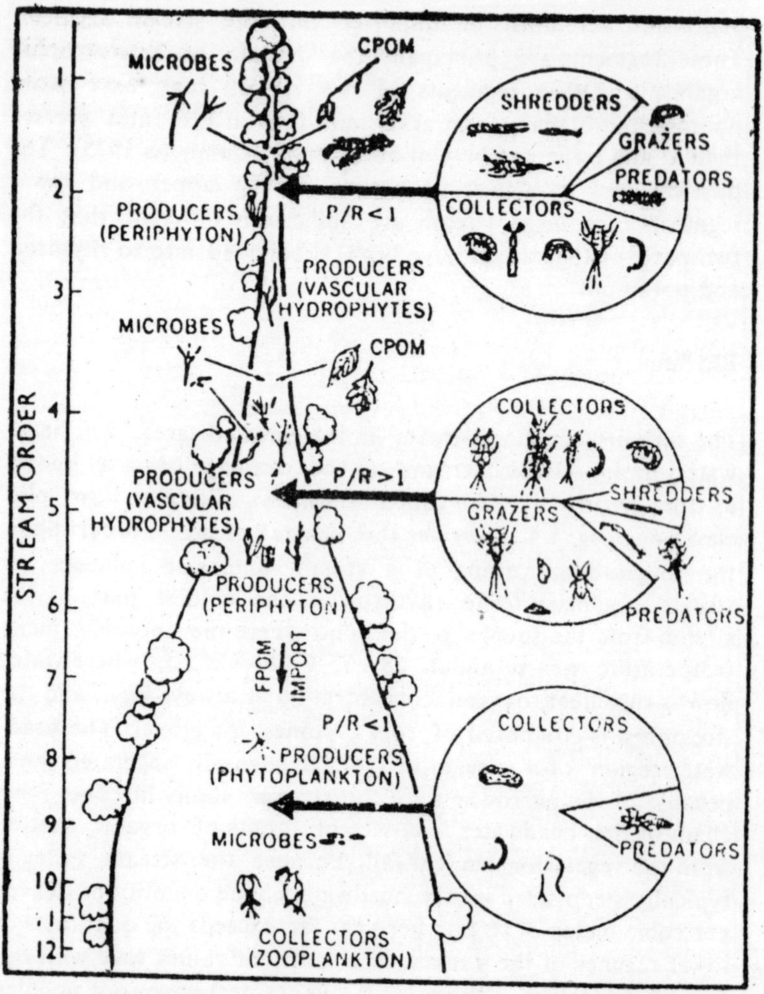

FIG. 1.4: The stream organism as a function of the stream order and position in the vertical stream profile.

becomes line particulate organic matter (FPOM) of the water and sediments. The shredders obtain their food source from microorganisms and hyphomycete fungi growing on leaf tissue (Williams 1981). One group of shredders that are important are aquatic hyphomycete fungi of the genus Ingoldia (=Gyoerffyella) which colonize leaves that fall into streams during the autumn (Barlocher 1982). Along with shredders are collectors which either trap the FPOM as it passes in the water or collect it from sediments. Caddisflies spin nets which capture suspended organic matter. The most ancient caddisfly taxa live in turbulent currents (Alstead 1982). Other organisms feed on the microorganisms that colonize the surfaces of organic particles and thus return the particles, minus microorganisms back to the FPOM pool to be utilized repeatedly by other benthic organisms down stream. Because of the fact that primary productivity is low, grazers are poorly represented. Predators such as stoneflies and fish species such as salmonids and sculpins are also present in these well-aerated, organically rich, high-protein rhythron waters.

Potamon

As one descends along the longitudinal profile of the stream, several physical and biological changes occur. The potamon (means river) is the lower stream segment where the monthly mean temperature rises to over 68-75°F (20-24°C), where oxygen deficits occur at times, where flow quickens and the channel deepens, and where the bed consists mainly of sand, silt, or clay. There is a gathering of water resulting from the confluence of additional streams, and the stream broadens to the extent that the middle becomes exposed to sunlight. In humid climates this happens when streams become-third-or fourth-order. The character of the stream shifts from a predominantly heterotrophic to an autotrophic condition. Planktonic forms are common in the surface waters, and in the benthic community, grazers and collectors predominate. Shredders become less common. The caddisflies of more modern taxa have evolved to live in the laminar currents which have greater concentrations of particulate food (Alstead 1982). Predators include both invertebrates and vertebrates. Collector species dominate the river benthos as a result of the inflow of large amounts of FPOM produced up stream, and locally produced FPOM from planktonic forms. Overall, the P/R ratio becomes less than 1 because high turbidity blocks sunlight. However, high turbidity may be a modern problem because many early accounts of streams along the east coast of North America indicated clarity where now there is turbidity (Trimble 1974).

The transition zone between the cooler, turbulent, shallow, and aerated rhythron and the warmer, deeper, smooth-flowing potamon generally has the largest number of species because of overlap in the ranges from both segments of the stream. The total number of fish tends to decrease while the weight per unit fish tends to increase in the downstream reaches of the stream. The weight per unit area remains more or less constant in the stream (Hynes 1970). North American rivers have many more species of fish than do northwest European rivars, though there are many similarities between the two funas. Glaciation had an unequal influence on the faunas of the two areas. In North America, as the climate changed, the river fauna was able to retreat to warmer climates in the south. In northwestern Europe, species were caught between the ice sheet building in the north and the glaciers forming in the Alps, resulting in the extermination of many species (Hynes 1970).

The above discussion is based on research conducted primarily in humid mid-latitude climates. There are many factors which influence the productivity of a stream. For example, the seasonality of precipitation and streamflow changes with such factors as altitude, latitude, continental position, that is, all of the variables that influence climate. These same factors, because of their role in productivity, also influence the amount of organic matter available in a stream. In addition, the composition of vegetation, whether coniferous or broadleaf deciduous forests (and even the species such as alder or oak), also affects the amount of organic matter in a stream. This brings the discussion back to the water balance idea discussed earlier in the chapter. A comprehensive view of the complexity of the stream biosystem requires combining the principal factors which govern the operation of the regional water balance with factors that influence litter production and organic matter availability in the stream. The average timing of streamflow, litterfall, and peak particulate organic matter transport into streams is shown for several North American regions.

The influence of latitude is equally profound. Middle-latitude streams include rhythronic and potamonic zones and support both cold and warm water adapted species. Potamonic species are capable of surviving—but not thriving—in rhythronic waters. Rhythronic animals cannot survive in the warm, oxygen-depleted waters of the potamon (Williams 1981). In high-latitude environments, the streams are typically rhythronic, supporting species adapted to cold, well-aerated water throughout the entire length. In tundra freshwater systems, organic matter has accumulated in the form of terrestrial peat, and is transferred to higher organisms via aquatic insect larvae (Schell 1983).

In lower latitude streams, the rhythron may be very short or absent, with the potamon dominating the river. Large tropical lowland rivars have distinctly high fish species diversity, and the fish are characterized by having many adaptations to low oxygen supply (Welcomme 1979). The Amazon undergoes seasonal fluctuations of as much as 20 m (Gentry and Lopez-Parodi 1980) and floods large areas during high water stages. The total efflux of organic carbon for the Amazon has provided a basis for estimating the total organic carbon input to the oceans from the Earth's rivers. The Amazon contributes about 10^{14} g of carbon and the total for the Earth's rivers must exceed 10^{15} g (Richey *et al.* 1980). One source of organic carbon is from swamp hardwood forests and floodplains. This organic carbon enters the river during flooding. The influx of organic carbon tends to be moderated by biological processes because the carbon promotes increased organic activity which in turn removes organic carbon from the river at a faster rate. The overall result is a relatively constant transport of organic carbon (Richey *et al.* 1980).

The floodwaters of the Amazon open up extensive habitats to aquatic organisms. During high water periods, the forested lowland floodplains are inundated, forming extensive varzea forests. These forests are the base for the aquatic food chain. Many common fishes move into the inundated forests and voraciously consume fruits and seeds. During low water periods, the same fish move back into the main river channel or floodplain lakes and consume little food (Goulding 1981). There are many fascinating examples that provide insight into the biogeographic patterns of streams and organisms.

Natural Impoundments

Several physical processes form water-holding depressions in the landscape. These include glacial action, stream meandering, tectonic forces, and structural weaknesses. During Pleistocene times, continental glaciers formed lake-pitted kame and kettle landscapes in many areas of the Northern Hemisphere, such as the lake areas surrounding the Great Lakes of North America or dotting the landscape in European countries around the Baltic Sea. Enormous blocks of ice were buried in glacial till and when they melted left depressions or kettles which filled with water, as well as ridges which are known as kames. Rivers play a role by creating oxbow lakes as they cut and fill their sedimentary banks and undergo course changes in their floodplains. Tectonic processes have formed the intermountain basins in western North America as well as coastal subsidence in the Atlantic and Gulf coasts of the continent. Structural weaknesses that have resulted from the weathering and dissolution of the underlying limestone formations have created sinkhole depressions in several regions, such as south Florida.

Rift Lakes

The rift valley lakes of East Africa are spectacular for their size and for the diversity of their fish fauna (Fryer & Iles 1972, Beadle 1981). Lake Malawi is the southern-most rift valley lake; it is 620 km long, 80 km wide and has a single large basin which is 750 m deep. Lake Tanganyika is even larger and deeper and has three basins, but has a smaller diversity of fish.

Lake Malawi has more, species of fish than any other lake in the world. About 300 species of fish are currently known but many new species are being described, and it is estimated that at least 1,000 species probably exist in the lake (Ribbink *et al.* 1983). The most fascinating biological feature about Lake Malawi is that nearly 95 per cent of the fish are members of a single family, Cichlidae.

Cichlids are a large family of freshwater fish and are found in Southern Africa, Sri Lanka, India, South America, and one species is found in Texas. Their distribution (with the exception of the member found in Texas) suggests a long history which extends back to a time when the continents of Gondwana were connected. All but four of the cichlids found in Lake Malawi are endemic to the lake. All species except one are mouth brooders. The exception is the red-bellied bream, Tilapia rendalli, which digs and broods in tunnels.

The density of cichlid species is truly amazing. On Domwe and Thumba Islands near Cape Maclear, in Lake Malawi National Park, there are rocky areas where densities of 15 to 20 different species of cichlids exist per 10 square metres (Reinthall & Lewis 1986). The fish have adopted a large number of specialized feeding behaviours. For example, algae and diatoms grow in the water column, on rocks and sand; different species have specialized their feeding behaviour to live in each of these conditions. In addition, some cichlids feed on the scales and fins of other fish. Lake Malawi is anoxic at about 230 m, and bacteria are the dominant organism in anoxic waters. In addition to nutrients sinking to depth and being lost, another way in which there is a loss of nutrients from the lake is through the hatching of midges (Chaoborus and Chironomus). So many midges hover in the air at one time that they form dark clouds above the surface of the lake. When the midges hatch and blow onto the land they carry valuable nutrients from the lake.

As in the case of the finches and turtles which Darwin found in the Galapagos Islands, the many related fish species existing within Lake Malawi raises questions regarding the development of new species.

Wetland Communities

One common successional hypothesis treats open bodies of water as ephemeral features of the landscape which over time fill in and disappear. Starting with an open body of water, there is a submergent stage (submergent water plants become dominant), an emergent stage (shallow water herbaceous plants dominate), and then "climaxes" with a closed stage of either a wet prairie or a wet forest. There are situations, such as within the floodplain, where this sequence may be appropriate. However, because of the qualities of the organic materials that have filled the depression, organisms may continue to play a role in maintenance of the body of water. There are a number of wetland communities of plants and animals which thrive within and are an inextricable component of the drainage system. A few of these communities, the bog, marsh, and swamp, are briefly described below.

Bogs

Two broad groups of soils are commonly recognized: mineral and organic soils. Organic soils are composed of large accumulations of mosses or sedges, and are commonly referred to as peat. They have the highest organic matter content of any soil, see Table 1.4 One of the outstanding physical properties of these materials is their high water-holding capacity. Mineral soils hold about 20 per cent to 40 per cent of their weight as water while undecayed moss or sedge peat can hold 12 to 20 times their dry weight as water. These materials are veritable sponges in the landscape.

Organic deposits of peat occur in many places on the Earth. They are located in structural basins, such as the Everglades of Florida, which has sedge-derived peat, but the major locations are in glaciated landscapes. Extensive deposits occur in northwestern Eurasia and North America.

TABLE 1.4

Comparison of Physical and Chemical Characteristics of Peat and Mineral Soils (Values in Per cent Based on Dry Matter, modified from Brady 1974)

Source	*Nature*	*Organic matter*	*N*	P_2O_5	K_2O	*CaO*
Michigan	Low lime	77.5	2.10	0.26	—	0.16
Michigan	High lime	85.1	2.08	0.25	—	6.80
Germany	Low lime	97.0	1.20	0.10	0.05	0.35
Germany	High lime	90.0	2.50	0.25	0.10	4.00
Austria	Low lime	93.3	1.40	0.10	0.06	0.45
Austria	High lime	83.8	2.10	0.18	0.13	2.38
New York	Low lime	94.2	1.26	0.15	0.10	0.60
New York	High lime	81.5	2.56	0.19	0.28	6.51
Minnesota	Low organic	59.7	2.35	0.36	0.17	2.52
Minnesota	High organic	94.0	1.70	0.16	0.04	0.31
Florida	Sawgrass peat	87.1	2.79	0.41	0.04	5.20
Canada	Peat soil	74.3	2.19	0.20	0.16	—
Washington	Woody sedge peat	89.2	3.52	0.43	0.09	1.29
Mineral soil (typical)		4.0	0.15	0.04	1.70	0.40

Sphagnum moss composes the peat of the bogs of large areas of North America and Eurasia. The water in these bogs is cold, anaerobic, high in dissolved organic matter, and low in minerals. The moss is essentially sterile, containing neither bacteria nor the organisms of decay that are usually associated with soil. The large bogs exert much influence on the hydrology of the surrounding region. They may cause either an increase or decrease on the annual spring flood peak, depending on how the bog is tied into the groundwater system. In general, the bogs reduce storm peaks significantly and augment streamflow during dry periods. Thus, rivers that are supplied by the bogs tend to have a more constant flow than other streams (Novitzki 1979). Bog communities thus exert a moderating influence on the stream system.

Marshes and Swamps

Marshes and swamps develop in moist climates, generally n low lying depressions, and play an important role in channelling the runoff of a stream system. If the depression is shallow and poorly drained, and supports plants and emerge above the water surface, it qualities as a marsh or swamp. When water enters the depression, the velocity decreases rapidly, and the organic matter, sedimentary particles, and dissolved minerals that have been carried by the stream flocculate and accumulate within a relatively small area. These small areas are the recipients of the nutrients and organic wealth of the watershed and are exceptionally rich places on the Earth.

Marshes and swamps are distinguished from each other by their vegetation. Vegetation varies with climate. The seasonal, continental mid-latitude climates support cattails and reeds that make up the marsh; warmer climates support the growth of trees such as cypress which make up the swamp vegetation.

Emergent marshes are extremely productive environments. They annually produce between 450 and 2,320 g/m^2 (dry weight) of above-ground material and compare favourably with saltwater marshes in their productivity (Good *et al.*, 1981). In order of their productivity are cattails (Typha), reed canary-grass (Phalaris arundinacea), reeds (Phargmites communis), sedges (Carex), and bulrush (Scirpus) (Pratt & Andrews 1981). Marshes can undergo enormous fluctuation of productivity from year to year. Prairie marshes have been described which vary by as much as 18-fold over a period of 5-20 years (Good *et al.*, 1978). Marsh variability is related to two major factors. First, weather fluctuation leads to changes in water depth that strongly influence the plants which can be supported within a body of water. During wet years, high water may flood out emergent plants, while dry years may result in the filling in of a marsh with emergent plants. Severe drought may even lead to the drying up of a prairie pothole, whence the unprotected bottom may become deepened by wind erosion. In North America an animal plays a major role in marsh vegetation, the muskrat (Ondatra zibethica). Muskrats are rodents that have an impressive capacity to reproduce. They have six or more young per litter and under optimal conditions, can have four litters per year. Muskrat populations have been reported to grow so rapidly that they literally "eat out" the plants from the marsh (Weller 1981).

The muskrats use the plants for food and construction. The rootstocks, rhizomes, and tips of the plants are eaten and the stems are used to construct a platform and left to rot. In the early stages of reestablishment in a marsh, muskrats tend to construct lodges nearly 2 m high and 4.5 m in diameter. However, the lodges are half that size when the muskrat population has removed a large proportion of the plant material. Reproduction rates have also been reported to undergo changes that reflect the conditions of the marsh. Thus, as the marsh becomes crowded, lodges decrease in size, and the number of surviving young decreases. These dynamic activities affect the numbers and species of birds which inhabit a marsh. The greatest number of individual birds are supported when the whole marsh is dominated by a thick growth of emergent plants. However, when there is a roughly equal balance between open water and emergent vegetation, then the maximum number of species can be supported by the marsh.

When muskrats open up the vegetation around their lodges, they create habitat diversity that attracts an increased diversity of bird species (Weller 1981).

The most extensive swamps in North America occur along the Atlantic and Gulf coastal plains. The most notable examples include the Great Dismal Swamp, Okefenokee Swamp, and Everglades. The Great Dismal Swamp consists of nearly 200 km^2 within the Atlantic coastal plain. The swamp is believed to be at least 80,000 years old (Kirk 1979). The oldest pollen is from pines and spruce. The modern vegetation consists of a variety of tree species including cypress, white cedar, and tupelo gum (Nyssa aquatica). The swamp acts to stabilize the groundwater table in the region. During periods of very heavy rains, the swamp controls floodwaters by temporarily storing excess runoff and releasing it slowly as runoff or into the groundwater system. The area abounds with an amazing variety of plants and animals, some of which are poorly known (Kirk 1979).

Within the Everglades of South Florida and the Gulf Coast swamps, one organism has had an especially important role within the environmental system, the American alligator (Alligator mississippiensis). Alligators are active diggers. They dig out channels and build large mounds for nest sites. They also dig deep pools or "gator holes" in the swamp. During periods of extreme drought these pools provide the last remaining inhabitable sites in the swamp for birds and other animals to survive. When the drought breaks repopulation of the swamp takes place from the animals which have survived at gator holes. The nesting mounds of alligators contribute to the variety of the swamp environment by providing nest sites for several bird species, such as herons and egrets. Thus, marshes and swamps are complicated habitats that depend on energy expenditure and behavioural adjustment of specialized water-dwelling animal inhabitants. The activities of these animals create conditions that provide opportunities for existence of the larger plant and animal community.

Beaver Impoundments

Beavers are a sine qua non of the influence of animals on streams. They have an unusually large impact on stream systems and the development of valley soils and vegetation of the Northern Hemisphere. Two species are recognized, the Eurasian beaver, Castor fibre, and the North American beaver, C. canadensis. The beaver is a large rodent, second in size to the capybara (Hydrochoerus hydrochaeris), a relative of the guinea-pig and inhabitant of tropical America. The beaver attains a length of between 90 and 120 cm and a weight of up to 30 kg. The present range of the beaver is only a fraction of its former range. In the Old World the range of C. fibre once extended from Scandinavia to Spain and from Britain to Siberia. Now there are only limited populations in the lower Rhone, the middle Elbe, and the southern part of Norway. The former range of C. canadensis included the northern half of North America with extensions into the southern Appalachians and the southern Rocky Mountains. They numbered in the thousands of millions and were an attraction to early traders who sought their pelts. In general, beavers were hunted for food (because of the scales on their tails they were defined as fish and could be consumed by Catholics on Friday and during Lent), for fur (they have magnificent coats which are durable and easily tailored), and for castoreum, a scent gland substance which was highly valued for its medicinal properties during medieval times. For these reasons and because they occupied flood plain soils that humans valued for agriculture, the populations of beavers were cut to a fraction of their former numbers by hunters and trappers. But, many soil and vegetation features can be understood only by considering the influence that beavers have on river valleys.

The activities of the beaver are entirely bound up with trees, flowing water, and the construction of dams. Beavers create deep ponds or maintain water levels in preexisting depressions by building dams across streams. These are usually made of logs and sticks, although dams made of boulders have also been described (Chapman and Chapman 1937). The impoundment stores water and serves the beaver by providing a site for a lodge which is safe from predators. The food of the beaver consists of the bark and shoots of poplar, willow, alder, elm, or oak, plus vegetation which is favoured by the expanded riparian environment that the beaver pond creates. The beaver obtains food by chopping down trees. The bark is stripped and the inedible portions of the trees are moved to the river for the construction of the beaver dam or the lodge which is built in the impounded waters behind the dam. In order to move large logs, beavers construct canals.

Castor canadensis is acknowledged as the superior dam builder. The largest recorded dam built by C. fiber was found in the Voronezh region of Russia. It was 120 m long, 1 m high, and 60-100 cm wide (Frisch 1974). One of the largest structures constructed by C. canadensis was on the Jefferson River in Montana; it was 700 m long and could bear the weight of horses and then riders (French 1974). The largest dams were probably maintained by generations of beavers for centuries. Beavers regulate the water level of the beaver pond by constantly adjusting the dam. During high water, the dam can be adjusted by lowering the crown. During low water the beavers tighten up the dam to reduce seepage. The splendid teamwork of beavers building and maintaining a single dam is itself magnificent. A series of dams within a number of tributaries of a drainage basin operate together to create an impact on the landscape.

The concerted efforts of many beavers must have changed many valleys of the Northern Hemisphere during the Holocene from relatively physically and biologically homogeneous habitats into heterogeneous environments supporting diverse groups of organisms. The process can be imagined within the high velleys of the Rocky Mountains. Prior to the movement of beavers into high valleys, conifers probably dominated the landscape. As beavers move in and construct dams, the impounded waters flood out the conifers. Each pond creates a site which is attractive to migrating waterfowl and other birds. The riparian environment of the shores of the pond supports new types of plants (poplar, willow, alder), that supply the food for the beaver population. The pond checks the movement of sediments down stream and these particles and to the richness for the aquatic organisms of the pond. The enlarged surface area of the enriched pool of water supports high levels of photosynthesis and releases increased amounts of organic carbon that in turn enhances the food supply for aquatic organisms down stream. Eventually, the impoundment becomes filled with sediments, and beavers abandon the site. As the beaver pond drains it becomes a thick lens of rich humus that supports the rapid growth of grasses and forbs (non-grass species) and establishes the beaver meadow. These soils have high water-holding capacity and slowly release water to the water table and the stream. The activity of the beaver becomes the primary force in the evolution of the soils in the riparian zone. They influence the diversity of habitats and establish structures which argument the slow but steady movement of water through the drainage basin. These feats give the beaver its reputation as the unparalleled engineer and builder of the animal kingdom (Hodgdon & Larson 1980).

The Estuary

Eventually, most freshwater streams flow to ocean level and meet the sea. Under special conditions, the flow of fresh water to the ocean is partially enclosed and the outlet to the sea restricted to a relatively narrow opening. These places are referred to as estuaries. The most important estuaries include drowned river valleys, such as Chesapeake Bay; down-faulted sections of the Earth's crust, such as San Francisco Bay; drowned glacial troughs, or fjords, found in Norway, Alaska, New Zealand, and other glaciated mountainous coastal regions: and saltmarshes and lagoons that are formed behind barrier beaches.

Within the estuary, fresh water from the land flows upon and meets the denser salt water of the ocean. The interface between the two different types of water is a nutrient-rich and highly productive site for organisms. But these sites are also physically demanding and only a few species can tolerate the rigors of successive exposure to fresh and sea water, and the wide fluctuation in temperature associated with these environments.

One component of the estuary is shallow saltmarshes that play an important role for organisms. Saltmarshes are regularly rinsed by the ebb and flow of the tides. Waste products are diluted and removed and nutrients are brought in from the land and the sea. The materials of streams flocculate as the mixture of clay-size particles and algae meets the sea waters. They settle out, leaving the water clear (Avnimelech *et al.*, 1982). In the saltmarsh, plants need not expend energy in collecting minerals. They are assisted in obtaining minerals from many organisms. Clams, oysters, and mussels trap nutrients as they feed, and deposit the wastes as pseudo-feces. In these nutrient-rich waters, plants grow, die, and cycle rapidly as they are consumed by the inhabitants of the estuary. The base of the food chain of the estuary is marsh grass or Spartina. The major energy flow is by way of the detritus food chain rather than the grazing food chain. In estuaries along the coast of Georgia, U.S.A., for example, Spartina alterniflora dominates, and organic detritus (over 90% of which originates with the growth of Spartina) is the link between primary production and consumption. Only a small portion of the grass is consumed as a part of the grazing chain (Ranwell 1972).

Estuaries are one of the most productive habitats on Earth. They provide huge food supplies as well as sanctuary for many species of waterfowl. They serve as a breeding ground and nursery for many marine fish species. They store nutrients for larval and juvenile marine life. As various species mature, they leave the estuary and head out to sea. Thus, there is a constant movement of the rich organic and nutrient materials to the sea (Honjo 1982). As tides move in and out there is also a movement of the materials out to sea. And thus it is that the stream flows to the sea and enriches the shallow coastal waters of the marine environment.

2

Freshwater Environments in Polar Regions

The chronic aridity of both polar heartlands has already been noted. Though the antarctic ice-cap is estimated to contain 90 per cent of the world's water, only a tiny fraction of it becomes available each year for living material. Similarly the arctic desert, with or without ice sheets, has very little water available during the short summers, and none in waters. On the polar fringes warmer air brings in moisture and deposits it as hoar-frost, snow or rain. Though water is solid for much of the year, the annual thaw in spring and summer provides a wide range of aquatic habitats that support biological activity, and may even allow high productivity for a few weeks each year.

Wet soil is essentially a freshwater habitat, above. Melting snow and ice surfaces provide transient substrates which bacteria and algae colonize, forming patches of pale green, red or yellow several hectares in extent. Snow-melt and rain gather in runnels, streams and even small rivers, that flow for a few hours, days or weeks in summer but are stilled by freezing for the rest of the year. Ponds and lakes form in poorly-drained rolling country, accumulating debris and nutrients, freezing partially or completely in winter but providing a limited habitat for plants and animals in summer.

Polar Freshwater Habitats

Snow and Ice Habitats

Freshly-fallen snow surfaces are virtually sterile, though bacteria are present in detectable quantities even on the southern polar ice-cap (Meyer-Rochow, 1979). Older snow surfaces become adulterated; damp snow attracts wind-blown particles electrostatically (Benninghoff and Benninghoff, 1985), acquiring films of rock dust, bacteria, spores and plant propagules. These accumulate as the snow melts under spring sunshine, and the resulting discoloration hastens the melting. Fragments of moss and algae that land on snow surfaces survive well in the moist environment, and may settle through the remaining snow to colonize the ground beneath.

Ionization and other atmospheric processes add nitrates and ammonium compounds to the snow in nutritionally significant quantities (Parker *et al.*, 1978). Snow close to the coast is often additionally contaminated by mineral salts and organic material from the sea. In summer these are carried inland in sea spray; in winter impregnated snow from the sea ice deposits them on coastal rocks and snowdrifts. In late spring, snow surfaces within a few hundreds metre of the coast often show a flush of colour, usually pink, green or brownish-yellow, caused by patches of unicellular or colonial algae. On the South Orkney Islands Chlamy-domonas, Raphidonema and possibly Ochromonas are implicated (Fogg. 1967), similar accumulations of algae in a wider range of species have been seen at many other points around the coast (Akiyama, 1979; Kol and Flint, 1968; for a recent review see Vincent, 1988). Their productivity is low, fixed carbon being of the order of 10 mg m^2 of snow per day; colour differences are due to different stages in development of the cells, and their sudden, often dramatic appearance during warm spells is more probably due to progressive loss of covering snow than to rapid proliferation (Fogg. 1967). Snow algae are widespread in alpine and polar regions, but require persistent banks of melting snow to form spectacular coloured patches; they are rarely seen, for example, on the colder coasts or interior ice-cap of Antarctica. Their ecology has been reviewed by Hoham (1975, 1980).

On the lower reaches of glaciers and on ice shelves close to the sea, where ablation exceeds snowfall in summer, wind-blown dust and rock fragments encourage local melting. The resulting shallow pits and pools, often extending over many hectares, accumulate salts and nutrients from year to year. Many are rich enough to support a distinctive cryoconite flora of algae and cynanobacteria, which survive winters encapsulated in the ice (Wharton *et al.*, 1981; Vincent, 1988) and flourish briefly the during the summer thaw.

Still-Water Habitats

Over much of the Arctic, surface drainage is poor; permafrost seals the subsoil, and snow-melt provides a flush of water in late spring. In the northern desert there is rarely enough snow for melt-water to accumulate, and ponds are rare. Where winter snow is plentiful, ponds and shallow lakes occupy large areas; over much of the tundra they form networks of static waterways that make walking difficult after the thaw. On ice-scoured uplands and coastal flats there may be more ponds and lakes than dry land. Coastal plains often feature long, shallow lakes, separated and aligned by raised beaches and occupying up to 90 per cent of the terrain. Antarctica has relatively little standing water. Large lakes, many of them strongly saline, are reported from several of the oasis areas; freshwater lakes are also present even in high latitudes, and temporary lakes form in summer alongside many glaciers. Small coastal summar ponds are rare on the mainland but plentiful in the maritime and peripheral Antarctic regions.

Lakes form also on poorly-drained ice, alongside glaciers and ice-caps, on snow-covered ice-sheets, and between ice-sheets and ice-filled moraines. Some are long-lived, fed by annual influxes of melt-water. Others are temporary, flooding during the summer melt and liable to sudden drainage through rifts in the ice and other causes (Hambrey, 1984).

Polar lakes and ponds are partly or completely frozen and snow-covered for much of the year. Large ponds and lakes, especially those in high latitudes, stay frozen all the year round, accumulating snow-melt water from rocks nearby and part-thawing only in exceptional seasons. Large lakes thaw round the edges but tend to retain a core of ice in summer; in consequence their waters never warm far above freezing point. Small lakes and ponds thaw completely. Lakes that owe their volume to a balance between inflowing and outflowing water may change radically between seasons, almost draining in winter when the inflow slows down, and overflowing when summer melt brings in water than cannot escape through still-frozen outlets.

Lakes that are ice-free in summer begin cooling at their surface late in the season. The cooled waters sink and circulate until the water mass reaches the critical temperature of 3.98°C, at which fresh water is densest. Thereafter cooler waters overlie warmer, and the surface freezes while bottom waters are still relatively warm. Shallow ponds freeze to the bottom in winter; deeper ones may remain unfrozen at depth, providing a still, protected habitat for overwintering fish, larvae, eggs and spores.

Conversely, spring warming begins immediately under the ice as soon as sunlight penetrates usually, in early summer after the surface snow (which is relatively opaque to solar radiation) has ablated or melted. Water that is warmed a degree or two above freezing point sinks into the colder water below, and is replaced; the resulting density currents stir the lake into turbulence and spread warmth, a process that is enhanced by wind-stirring as the ice disperses. Thus the water quickly becomes isothermal, at temperatures close to those of the air or ground surface nearby. Some stratified hypersaline lakes are permanently warmer in their lower layers than at their surface, sometimes because of volcanic heating, more often due to solar warming through the ice and upper layers of water.

Most polar lakes are highly oligotrophic, with clear waters and a very limited range of species inhabiting them. Notable exceptions are arctic oxbow and other remnant lakes formed on river plains, which may be muddy and well endowed with minerals. Ponds range from oligotrophic to highly eutrophic, according to their situation and history. The poorest are the clear waters of ice-scoured uplands, where there is little soil or vegetation. Those on low-lying land invested by tundra vegetation are usually richer, with a seasonal turnover of minerals and slightly higher productivity. However, such important nutrients as nitrates and phosphates are generally scarce, because rates of bacterial decomposition in the soils of the catchment areas are low, and very little leaches out into the water.

Ponds and lakes that are frequented by birds often have an enhanced productivity, based on nutrients from the birds droppings and decaying feathers. Even the richest polar ponds and lakes have fewer species and lower productivity than most ponds of similar size and shape in temperate latitudes.

A few polar lakes are saline, their salts originating either from sea water and spray or from rock weathering in their own catchment areas. Sea water on chilling precipitates salts sequentially, first calcium carbonate, then sodium sulphate at —8.2°C, sodium chloride at—22.9°C, and magnesium and potassium chlorides at —36°C. leaving a residual solution rich in calcium chloride (Nelson and Thompson, 1954; Thompson and Nelson, 1956). Deposits of many of these minerals, notably halites (hydrated sodium chloride) and mirabilite (hydrated sodium sulphate) occur in the sediments and surrounds of many existing lakes, and in dried-out lake deposits. Saline water reaches its maximum density at temperatures below 3.98°C: at 27ppt and above the maximum is reached at freezing point, which is also determined by salinity.

Running-water Habitats

Few rivers start in polar regions, though some very large ones flow across arctic and subarctic lands to discharge into the Arctic Ocean. The great northwards-flowing rivers of Siberia and the lesser ones of North America (Figures 2.1) bring year-round flows of water and warmth from the south. They bear thick ice in winter and their flow is strongly seasonal (Table 2.1), especially so in those that rise or have much of their catchment area in tundra or boreal forest. Over 35 per cent of the annual flow of the Lena, in Siberia, is delivered during the peak month of June, less than 1 per cent in each of the three months of February, March and April.

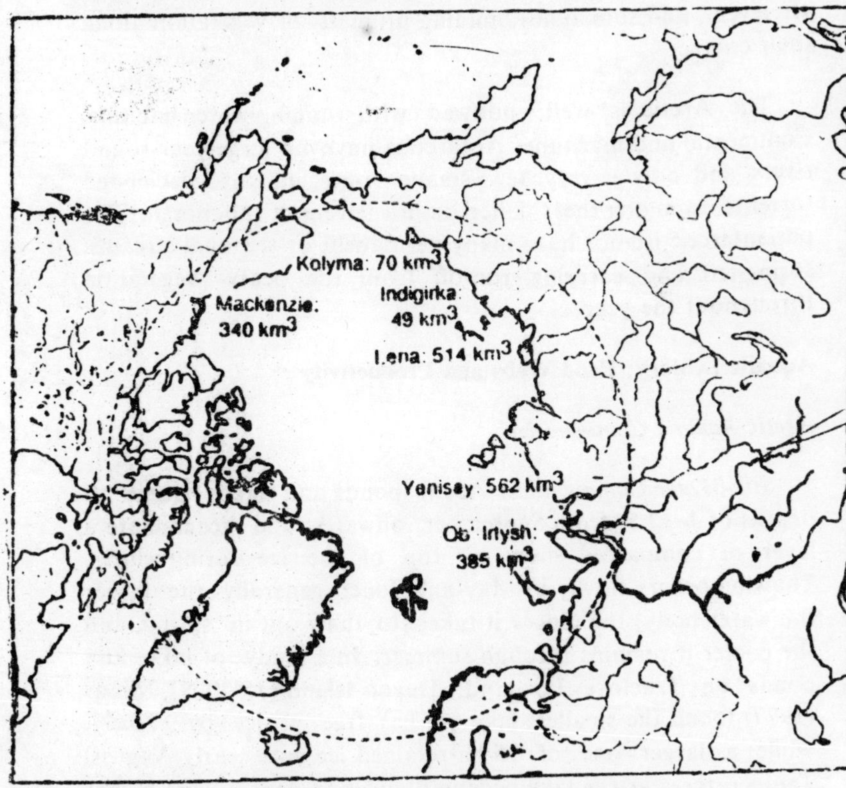

FIG. 2.1: Major inputs of fresh water into the Arctic Ocean: data from Table 0 1.

More characteristic of the regions are small graded rivers and streams fed by springs and surface runoff, and seasonal, short-lived torrents that flow from glaciers. In these up to half the annual output may occur in the three or four weeks of springtime thaw. Like the larger rivers they are erosive, undercutting and destroying their banks to form wide, braided channels. They carry heavy loads of sediment, abandoning it along their course and inextensive deltas; the finer deposits are often colonized by vegetation, providing some of the richest and most productive communities in the tundra mosaics. More characteristic still are the thousands of tiny meltwater streams that trickle from ice-sheets and snow-banks for a few days or weeks each summer, often following the same channels year after year, and sometimes building up mats of vegetation along their course.

The Arctic is well endowed with running-water habitats. Continental and maritime Antarctica have no large year-round rivers, and only a very few smaller ones, mostly glacier-fed torrents, to match the smaller tundra rivers of the north. The periantarctic islands have many permanent or seasonal streams, spring-fed and carrying run-off from the peaty vegetation throughout the year.

Aquatic Ecology: Food Webs and Productivity

Arctic Aquatic Communities

Still-Water Communities. Arctic ponds and lakes freeze to depths of 1 – 2.5 from September onward, and accumulate a layer of compacted snow on top of the ice during winter. Thawing occurs slowly in May and June; generally the larger the water body, the longer it takes to thaw out in spring, and the cooler it remains through summer. In a study of lakes and ponds on Truelove Lowland, Devon Island (75°30′N), Minns (1977) found the smallest lake (16 ha) free of ice by 17 July, while a larger one of 95 ha retained ice until early August. Temperatures at the lake-beds in about 6 m depths were higher than at the surface when ice was present, rising from 1°C in mid-May to about 3°C in late July. When most of the ice had gone, wind stirred the waters; by late summer temperatures became a uniform 4 – 5°C at all depths. Temperatures of nearby ponds rose higher, to 8°C.

Char Lake (72°42′N), the subject of long-term studies on Cornwallis Island, with an area of 52.6 ha, was typically ice-free from early August to mid-September. The waters were layered from October to May while ice was present, with temperatures close to freezing point near the surface but at 1 – 1.25°C near the bottom in 26 m. From June onward warming through the ice set up circulation currents which gradually raised temperatures to a uniform 4°C throughout the water column (Rigler, 1974).

Biological activity starts long before the ice has melted. In Truelove Lowland lakes aquatic algae and bacteria begin photosynthesis as early as February, when snow still mentles the ice and less than 1 per cent of available light penetrates (Hobbie, 1962, 1964; Holmgren, 1968). Often the algae accumulate in a thin layer on the underside of the ice sheet, dispersing when the ice melts. Measuring chlorophyll a in water samples as an index of phytoplankton abundance, Minns (1977) found it already plentiful under the ice by mid-May and reaching peak values in late May and June; Kalff *et al.* (1972), working on Char Lake and a neighbouring lake in the Canadian Arctic, recorded active production from February onward.

Algae that are capable of photosynthesis at very low levels of light intensity proliferate early in the season, to be replaced by a succession of other species as light intensity increases (Kalff, 1970). Primary productivity is generally low, averaging 10 – 50 mg Cm^{-2} daily in May and June and rising to 70 – 80 mg Cm^{-2} daily during the summer peak of production; on average only 6 – 12 g Cm^{-2} fixed throughout the year (Kalff, 1970; Minns, 1977).

Lack of nutrients, rather than low temperature or incident radiation, is the single most important factor restricting

TABLE 2.1

Annual and Quarterly Flows of Six Major River Systems into the Arctic Ocean. After Mackay and Loken (1974)

	June—Aug.	*Sep.—Nov.*	*Dec.—Feb.*	*Mar—May*	*Annual flow (km³)*
Ob'-Irtysh	57	20	9	14	385
Yenisey	57	18	7	18	562
Lena	71	22	4	3	514
Indigirka	85	14	<1	1	49
Kolyma	73	17	2	8	70
Mackenzie	38	27	19	16	340

primary production. Production is often highest in early summer when nutrient levels are maximal, though the water is still close to freezing point. By the time the water warms, all available nutrients have been absorbed and productivity falls. Only a few lakes show a second, late-season peak of production. Shallow lakes sometimes have a bottom layer of aquatic mosses, which by photosynthesis add oxygen to the water close by. Deeper lakes are likely to have the bottom sediment-covered, and the lower water layers may remain unstirred and stagnant.

The fauna of these lakes is usually restricted to rotifers, tardigrades and other small detritus-feeding invertebrates, a few species of copepods and other crustaceans, enchytraeid worms and the larvae of chironomid flies. The planktonic copepods Limnocalanus macrurus produce eggs from September to November; the nauplii which hatch after about one month mature in the following June (Roff, 1972, 1973; Roff and Carter, 1972). Mysids Mysis relicta by contrast have a two-yearly cycle; females require the same amount of energy for breeding as individuals of the same species in warmer climates, but take a year longer to accumulate it (Lasenby and Langford, 1972, 1973). Chironomids; including several species which bite man and other warm-blooded animals, are plentiful in the lake, forming the main food of arctic char, the single species of fish. Faunal relationships in Char Lake are reviewed by Rigler (1974): studies of individual species or groups include those of Morgan and Kalff (1972) on bacteria. Nurminen (1973) on oligochaetes, Welch (1973) on chironomid larvae and Holeton (1973) on arctic char.

Hypersaline lakes are fairly common in coastal areas of the Arctic which have until recently been ice covered, and since liberation have risen isostatically from below sea level (Figure 2.2); see Hattersley-Smith and others (1970). On Garrow Lake, a hypersaline lake of 418 ha on Little Cornwallis Island, northern Canada (75°N) Dickman and Ouelett (1987) found an ice cover up to 2.4 m thick lasting almost 11 months of the year. With a maximum depth of 50 m and little opportunity

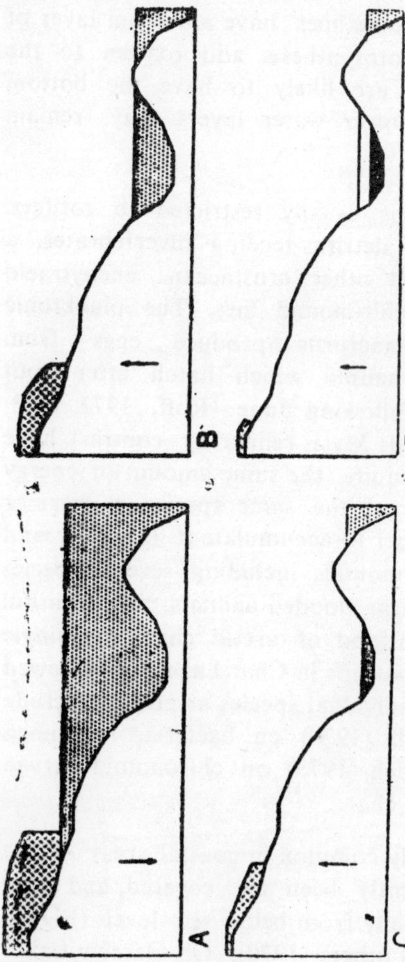

FIG. 2.2: Development of saline lakes on polar coasts. After depression by late Quaternary ice caps (A) the land rises, isolating hollows and arms of the sea (B). In dry areas, salinity of the isolated water rises through evaporation and other causes (C), producing (D) a dense bottom layer that subsequent inflows of fresh water fail to disturb. After Dickman and Ouellet (1987).

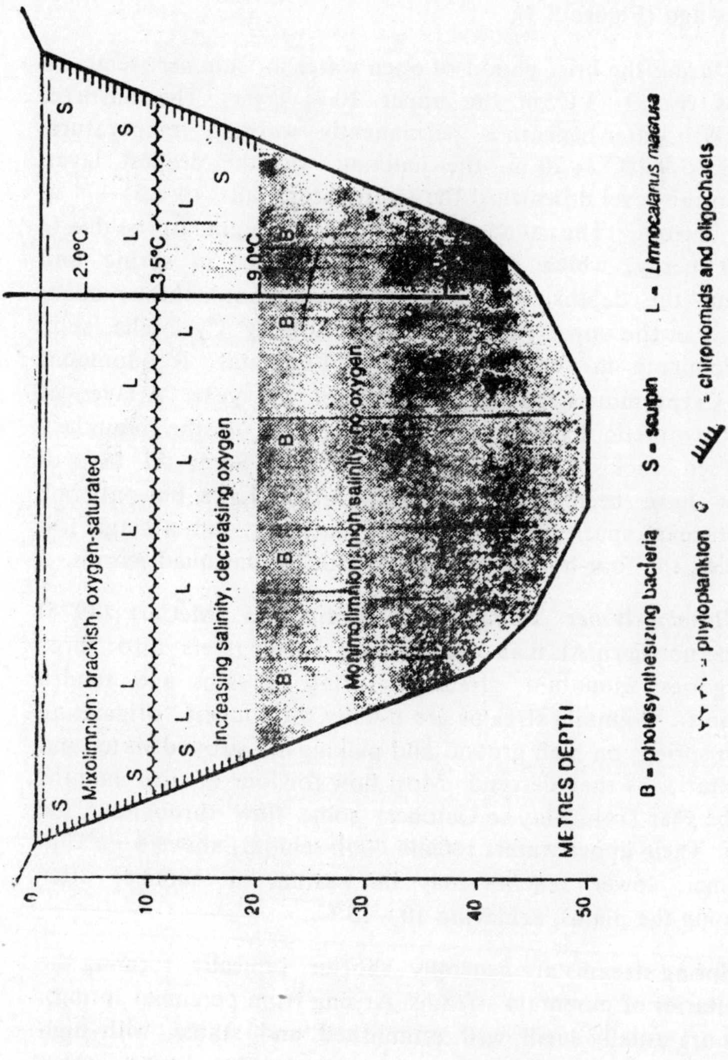

FIG. 2.3: Schematic diagram of life and environmental conditions in Garrow Lake, a meromietic saline lake of the high Arctic. After Dickman and Ouellet (1987).

for wind-stirring, the lake waters are strongly layered. A brackish upper layer of about 10 m is underlain by about the same depth of more saline water; beneath from about 20 m lies a permanently stagnant bottom layer of ancient sea water—a memento of the island's emergence from the sea some 3000 years ago (Figure 2.3).

During the brief period of open water in summer temperatures rise 2 – 3°C in the upper 10 m layer. The unstirred brackish water beneath is permanently warmer, temperatures rising to 9.1°C at 20 m, the junction with the deepest layer. From this level downward the temperature falls to 6.5° – 7° at the bottom. The anomalous heat is thought to be due to solar energy, which penetrates the surface ice in spring and warms the depths. Phytoplankton and zooplankton occur mainly in the upper 10 m. Diatoms, notably Cyclotella spp., predominate in summer, and microflagellates Rhodomonas and Cryptomonas spp. for the rest of the year. A layer of photosynthetic bacteria lives permanently on the boundary between brackish and highly saline water. Some 51 taxa of algae have been recorded from Garrow Lake, but only one crustacean species Limnocalanus macrura, and a single fish species, the four-horned sculpin Hyoxocephalus quadricornis.

Running-Water Communities. Craig and McCart (1975) divide northern Alaskan streams and small rivers into three categories; mountain streams, spring streams and tundra streams. Mountain streams are usually the longest, originating from springs on high ground and picking up ground water and tributaries as they descend. Most flow for four or five months of the year from May to October; some flow throughout the year. Their upper waters remain cool, seldom above 4 – 5°C in summer; lower reaches may be warmer in summer after crossing the plains, achieving 10 – 15°C.

Spring streams are generally shorter, typically forming the tributaries of mountain streams. Arising from perennial springs, they are usually small, well established and stable, with high mineral content; many flow year-round. The spring water temperatures remain low and fairly constant throughout the year, seldom failing to freezing point in winter or rising above 8-11° in summer. Tundra streams flow in meandering channels through tundra vegetation, draining the peaty soils, picking up humic acids and acquiring a low pH as they go. Frozen solid in winter, they warm quickly in spring sunshine and may reach temperatures of 15-16° C by high summer, but are the first to slow down and freeze in autumn.

Of the three kinds, mountain streams are least productive, with the smallest standing crop of invertebrates and least variety of species. Tundra streams are richer in invertebrates by a factor of 10, spring streams richer still by a factor of 100. Mountain and spring streams support populations of arctic char Salvelinus alpinus; tundra streams support grayling Thymallus arcticus, which spawn in their sluggish waters.

Antarctic Aquatic Communities

Still-water communities. Antarctic freshwater ecosystems include permanently frozen lakes high among the inland mountains, seasonal inland or near-coastal lakes of the ice-free areas, some of which are highly saline, and fresh or partly-saline coastal lakes and ponds (Heywood, 1977, 1984). Priddle and Heywood (1980) regard antarctic lakes as forming an evolutionary series which has developed over the period of about 20000 years since the ice-sheet was maximal; this is in any case a convenient way of describing them.

Proglacil lakes. The simplest lakes lie on or in front of ice-sheets, formed by seasonal melting that may be enhanced by the presence of scree or windblown dust. In Antarctica many such lakes occur inland, some on shelf ice close to sea level, others high among mountains or nunataks. Of hundreds recorded from the air, only a very few have been visited even briefly by sledging parties, and little is known of their biology. They accumulate from annual melt-water off the mountains and ice-sheets; their temperatures remain close to freezing point during a superficial summer thaw, though some of the deeper ones may have a permanently unfrozen core. Nutrients are extremely limited, and the lakes support only thin populations of algae and bacteria.

Lakes formed in rock basins by retreating ice-sheets pass through several evolutionary phases. As a moraine-dammed valley is exposed, water flowing in seasonal torrents accumulates on the valley floor, forming a lake that lengthens and broadens as the ice face retreats. The presence of a lake shows the inflow of water, causing rock debris and finely-ground rock flour from the base of the melting ice-sheet to be deposited as sediment on the lake-bed. The surrounding catchment area at first contributes little but clear run-off, and the lake water remains oligotrophic for a long period. However, as vegetation and mineralized soils develop in the catchment, minerals, organic nutrients and soil microorganisms begin to enter the lake from this source. Lakes close to the sea acquire minerals faster than those inland, from wind-blown spray and from seals, penguins, gulls and skuas, which like to wash in fresh water.

Adamson and Pickard (1986) describe several kinds of freshwater lakes in the Vestfold Hills oasis area. Pelite and Chelnock lakes, respectively supraglacial and proglacial, are fed by melting ice cliffs; Crooked Lake is an open-valley lake fed by a through-flowing river and there are many closed-valley lakes reflecting the water balance prevailing over previous centuries; one, Thalatine Lake, has over 40 minor beach terraces reflecting some 1300 years of evaporation. Others, for example, Watts Lake, began as marine inlets but were isolated by isostatic uplift and washed free of salt water by melting land ice; there are also saline lakes of similar origin (see below). Many other antarctic lakes are know at these early stages of development. Few on mainland Antarctica gain substantial nutrients from the soils of their catchment, except around the coast where birds and seals become important agents of eutrophication (Goldman, 1970).

More advanced lakes are found in the maritime Antarctic. Long-term studies of 17 lakes on Signy Island (Heywood *et al.*,

1980; see also Hawes, 1985) provide the most thorough coverage of a wide variety of lake types. Heywood (1984) summarizes the ecology of three lakes of varying nutrient content, occupying valleys of a coastal lowland. These typify many other lakes of similar size and aspect throughout the maritime Antarctic (Heywood, 1987). For studies of similar lakes on the South Shetland Islands see Paggi (1987) and Kieffer and Copes (1987).

Formed in ice-cut cirques or shallow, steep-sided valleys, the Signy Island lakes are moraine-dammed and often rimmed by a rubble-covered shelf about 1 m below the surface; the shelves are swept clear of sediment by wind-induced turbulence, and fall away to a narrow, sediment-filled trough. There is no emergent vegetation. In an environment of heavy snowfall, where mean monthly temperatures are above freezing point in summer but down to—15° C in winter, the lakes are frozen over with 1-2 m of snow-covered ice for 8-11 months each year. Temperatures at the bottom, down to 15 m in the deepest, fall to about 1° C in winter. In summer when melting is well advanced, persistent winds stir the waters and the lakes become isothermal, with temperatures up to 6° C. During this period they receive varying quantities of nutrients and bacteria (Ellis-Evans and Wynne-Williams, 1985) and yeasts (Ellis-Evans, 1985a and b), washed in from their catchments (Table 2.2).

Oligotrophic lakes. Moss Lake is typical of oligotrophic, high-level lakes, recently formed and remote from the sea. In a glacial cirque basin 48 m above sea level and 800 m from the shore, it has a surface area of 15000 m^2 and a mean depth of 3.4 m. Most water and ions enter the lake directly from precipitation or snow-melt. There is little leaching from the catchment area, which is small and covered with perennial ice, bare rock or scree, with hardly any mineral soil or vegetation. The waters are clear, and salinity and nutrient levels are very low (Table 2.2).

Phytoplankton of diatoms and other green algae is poorly developed; chlorophyll a values of up to 8.0 $\mu g\ l^{-1}$ were

TABLE 2.2

Salinity and Nutrients of three Signy Island Lakes: S=Summer, W=Winter, Data from Heywood (1984)

	Moss *S*	*Lake* *W*	*Heywood* *S*	*Lake* *W*	*Amos* *S*	*Lake* *W*
Chlorinity (mg 1^{-1})	23.0	49.0	35.0	56.0	37.1	94.7
Nitrite N (μg 1^{-1})	1.0	4.1	5.3	21.0	204.0	85.0
Nitrate N (μg 1^{-1})	21.0	50.0	39.0	128.0	321.0	721.0
Orthophosphate P (μg 1^{-1})	0.5	0.4	15.9	58.0	597.0	260.8
Silicate Si (μg 1^{-1})	143.3	267.0	174.0	368.0	191.6	656.9

recorded, and annual production was estimated at 12.6 g C m^{-2} in a year (Ellis-Evans, 1981a). In compensation, these clear lakes often have a rich benthic flora of perennial mosses Calliergon, Drepanocladus spp.) and algae (for example Tolypothrix, Plectonema and Phormidium), both liberally covered with epiphytic blue-green and green algae. Growing slowly to form dense mats, some with strands several m long, these fix up to 6 g C m^{-2} annually. The calanoid copepod Pseudoboekella poppei is a planktonic browser, fed on in all stages of its development by the larger copepod Parabroteas sarsi. On the lake-bed cladocerans Macrothrix hirsuticorna, Aloma rectangula and Ilyocryptus brevidentatus and two ostracods Notiocypridopsis frigogena and Eucypris fontana browse or scavenge on the benthic mats, which contain a microfauna of protozoa, turbellaria, rotifers, tardigrades, nematodes, gastrotrichs and enchytraeid annelids (Heywood and others 1979, Heywood, 1984).

Mesotrophic lakes. Heywood Lake typifies older, more mature lakes closer to the sea. It lies in two adjoining basins 4 m above sea level and 200 m from the shore, with a surface area of 45000 m^2 and a mean depth of 2.0 m. An extensive catchment area with minerals soils, frozen in winther but waterlogged in summer, supports swards of moss and lichens which contribute minerals to the water (Table 2.2); the lake is also visited by seals in summer. Waters are clear under the ice in winter but turbid from organic and mineral particles when the ice has melted.

Salinity and nutrient values are very much higher than those of Moss Lake, and the waters are markedly more productive. Chlorophyll a values up to 170 mg m^{-2} were recorded in spring and up to 40 mg m^{-2} in summer. Over 3 g C m^{-2} were fixed daily during the summer peak, and the annual rate of production was estimated at 173.0 g C m ^{2} (Light *et al.*, 1981) or 139-270 g C m^2 (Priddle *et al.* 1986). Lakes that are rich in phytoplankton and other suspended matter tend to have correspondingly less benthic vegetation, presumably because of light limitation in summer. Bacteria were plentiful in the water column of Heywood Lake; Hawes (1985) provides evidence that they enhance algal photosynthesis in lakes where nutrients are low, by speeding the recycling of carbon, nitrogen and phosphates. Planktonic crustaceans include Pseudoboeckella poppei, Parabroteas sarsi and immature stages of the fairy-shrimp Branchinecta gaini; mature stages tend to browse deeper on the benthos. There are fewer invertebrate species in the benthic algal mats, which are less developed than in Moss Lake.

Eutrophic Lakes. Amos Lake lies 10 m above sea level, adjacent to the shore on the windward side of the island; like many others in this situation, it is vastly enriched by marine influences. A small lake of surface area 6000 m and mean depth 1.7 m, it has a relatively small surrounding catchment basin, but the area is a popular summer resort for seals and nesting giant petrels Macronectes giganteus. There is little vegetation, but excreta and moulted hair contribute organic material to the soils, and the waters of Amos Lake are saline, turbid and nutrient-rich (Table 2.2).

Well stirred by seals, Amos Lake was too marky for direct productivity comparisons with the other lakes. Productivity was concentrated close to the surface; chlorophyll a values were high in summer and a daily peak production of over 300 mg cm^{-2} was recorded, drecreasing rapidly below 1 m (Heywood, 1984). Pseudoboekella poppei, Parabroteas sarsi and Branchinecta gaini were present in the water column, but there was little or no benthic vegetation or associated microfauna. Bacterial counts were generally high. Both aerobic and anaerobic bacterial decomposition contributed to nutrient recycling in lakes at all nutrient levels (Ellis-Evans, 1982; Hawes, 1985); the role of anaerobic bacteria is enhanced in winter when the water is static under its ice cover. Where nutrients are plentiful, productivity is high, production far outstrips decomposition, and organic sediments accumulate on the lake beds.

Saline Lakes. Lakes with densely saline waters are found in several of Antarctica's oasis areas. Their origins and histories are varied, but all occupy ground that was previously glaciated and many are ancient. Among the largest and most fully-studied lake systems, Lake Bonney in the Dry Valleys of South Victoria Land has probably existed for over 100,000 years (Hendy *et al.*, 1977), and a sediment core 2.8 m long from neighbouring Lake Fryxell is estimated from C-14 dating to represent over 50,000 years' deposition (Parker and Simmons, 1985). Similar lakes are known also in smaller coastal oases at Bunger Hills, Vestfold Hills, Schirmacher, Soya and Thala Hills. The ecology of saline lakes has been reviewed by Heywood (1984) and Vincent (1988).

Saline lakes of the Vestfold Hills oasis (Figure 2.4) string out along a valley, formerly a marine inlet, which rose and was cut off from the sea 5000-8000 years ago when the local land ice melted (Kerry *et al.*, 1977). Their salinities vary according to how long ago their waters were left to evaporate, and the amount of fresh water entering them annually; of the four main lakes, Deep Lake, the deepest (34 m) and most salty, has a mean salinity of approximately 280 ppt, roughly eight times that of sea water and similar in density to Dead Sea water. Deep Lake, and the slightly less salty Club Lake, seldom freeze over, though their surface temperatures drop almost to −20°C in winter. Wind-stirred in all but the deepest layers throughout winter, in summer they become thermally stratified, with surface temperatures over 10°C. Algae and bacteria are present in all the lakes at all depths; productivity is very low, and few herbivores or scavengers have been recorded. Metazoans are virtually absent except in some of the lakes where freshwater seeps in at the edges.

In higher latitudes, the saline lakes and ponds of South Victoria Land present a wide spectrum. Many are well above sea level and appear to owe their high salinity entirely to long-term accumulation and evaporation (for a discussion of this point see Heywood, 1984). Don Juan Pond, 122 m above sea level in Wright Valley and now only a few centimetres deep, is assumed to be the remnant of a much larger freshwater lake of

Lake	Area km^2	Greatest depth m	Surface below s.l. m	Volumes x10^6 m^3 present	Volumes x10^6 m^3 former	Volume ratio
Dingle	0.68	15	8.8	4.0	18.5	4.6
Stinear	0.83	18	12.0	5.4	26.0	4.8
Deep	0.64	36	50.4	9.1	60.7	6.7
Club	0.91	?	36.0	?	?	?

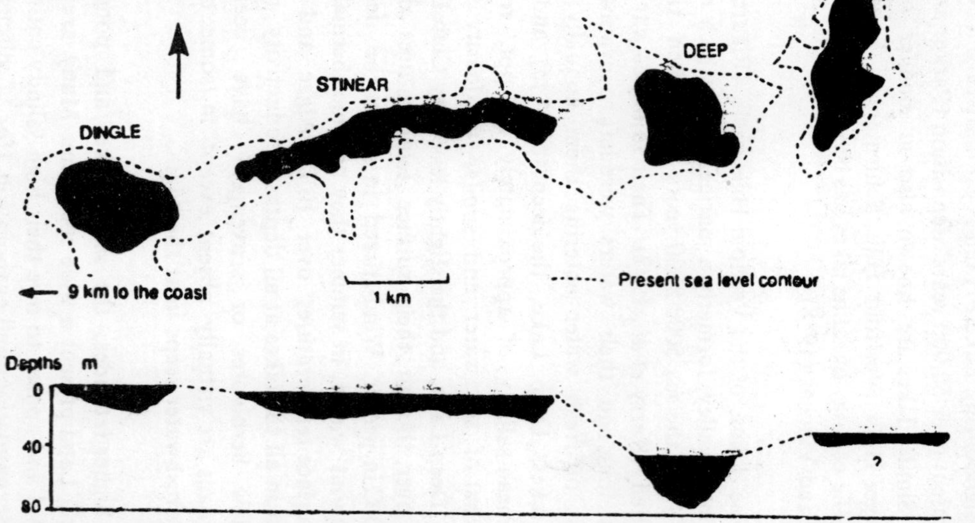

FIG. 2.4: The Vestfold Hills lakes system (data from Kerry and others, 1977). Reduction in volume since the land uplifted has resulted in salinities up to eight times that of seawater.

at least 10 m depth. Currently fed by a saline stream and limited snow-melt, its salinity, dominated by calcium chloride, is about 13 times that of sea water, and high enough to ensure that it seldom if ever freezes. The pond is virtually sterile except for algal mats which grow only at points about its edges where fresh water is available (Harris *et al.*, 1979; Wright and Burton, 1981).

Lake Vanda, also in Wright Valley, is a highly-stratified lake some 5 km² in area and up to 70 m deep. The remnant of a larger lake that once filled the surrounding valley, it has a superficial layer of fresh or slighly saline water down to 50 m deep, replenished seasonally by glacial streams, notably Onyx River, and stirred by their turbulence. Beneath lies a core of ancient salty water with a concentration of 140 ppt. and a temperature of 25°C; the origin of this warmth is uncertain, but more likely to be solar than geothermal. The lake surface is permanently frozen to a depth of 2–4 m, with ice that thaws only around the edges in summer. Several other lakes in the area share some or all of these characteristics.

The ice cover inhibits wind-induced turbulence, minimizes penetration of sunlight and keeps the temperature of the surface waters almost constant. Algae are present in the upper layers; productivity is low even in summer, and probably restricted to a shallow near-surface layer beneath the ice. Oxygen produced by photosynthesis accumulates in surface waters, but the saline substratum is anoxic; nitrifying bacteria accumulate at the boundary between, forming a substrate where algae and heterotrophic bacteria also congregate, and photosynthesis is maximal. Sulphur bacteria may be present below the anoxic zone, reducing sulphate ions and liberating hydrogen sulphide. Benthos around the lake edge includes mats of blue-green algae and moss, often with a microfauna of rotifers, tardigrades and other scavengers.

In the productivity of Lake Vanda phosphorus is a limiting element (Vincent and Vincent, 1982) because of gravitational losses into the sediment and lack of turbulence to return it to general circulation. Phosphorus that diffuses upward from the bottom is absorbed by the accumulated bacteria and algal cells at the top of the anoxic layer; very little penetrates to the less saline layers above, which are thus limited to the very small amounts brought in by Onyx River. For a full discussion of nutrient cycling in this remarkable lake see Vincent (1988).

Running-Water Communities. Antarctica has many small seasonal streams, and a few large enough to be called rivers. Nearly all are coastal; those of the continent are fed mainly from melting glaciers, those of the maritime region mainly from melting snow. Usually dry throughout winter, they start with the onset of spring thaw. Unbuffered by variations in gradient, their flows often fluctuate widely from afternoon maxima to early morning minima. Their waters seldom have opportunities to spread and be warmed: temperatures are usually less than 5°C. Continental streams run for one to three months annually, those of the maritime Antarctic for four to six months or more. Many carry finely-divided glacial sediment.

Plankton in Antarctic streams is usually low, restricted to algae or other organisms washed from lakes or ponds along the route. Stream beds are sometimes more productive, those with clear water supporting films or mats of cyanobacteria (Phormidium, Nostoc. Oscillatoria) and epilithic algae (Binuclearia, Prasiola). Beds of continental streams dry out quickly after the brief summer flow; any ice present ablates, and the desiccated mat is subject to intense winter cold. Those of the maritime region remain moist for a longer season, and usually remain protected by ice or snow, if not by water, throughout winter. In consequence they support richer algal mats, often with well-developed filamentous species (for example Klebsormidium, Mougeotia) which are active for much of the year (Vincent, 1988). Though they appear green and contain abundant chlorophyll, productivity of these mats is generally low: much of their material is moribund, and persists only because of an absence of decomposers.

Though the periantarctic islands are well endowed with ponds, lakes and streams, comparatively little is known of their biology.

Summary and Conclusions

Though water is a solid for much of the polar year, and generally poor in minerals even when in liquid form, both snow and aquatic polar habitats support life. Melting coastal snow supports an ephemeral algal flora that stains it red or green. Polar ponds and lakes are limited in their energy uptake by snow-covered ice, which also inhibits wind-stirring for much of the year. Permanently frozen catchment areas limit the amount of minerals available, though there is usually a small intake from soils during the summer thaw. Because food chains are simple, minerals absorbed into primary production settle in sediments where, in the absence of decomposers and detritus feeders, they may be lost to circulation. Some polar lakes, especially deep ones close to the coast and those in snow-free areas, may be saline with strongly-developed layering. Productivity is generally low in both still-and running-water habitats, limited mainly by lack of nutrients.

3

Biotic Factors of Freshwater Environment

FRESHWATER HABITAT

Freshwater environments are also divided into two major categories, lotic (lotus=washed, or running-water), and lentic (lenis=calm, or standing-water) habitats. Lotic habitats are those existing in relatively fast-running streams, springs, rivers and brooks. Lentic habitats are represented by lakes, ponds, swamps, etc.

The above classification of the freshwater environments is based on two conditions: currents and the radio of the depth to surface area. Since lakes and ponds often contain currents or at least wave action and since streams often harbour quiet pools or calm backwaters' the difference between lotic and lentic waters is not very precise. However, temperature, light, currents, concentration of respiratory gases, and concentration of biogenic salts are important limiting factors influencing organisms of all fresh water habitats.

Lentic Community

Lentic waters are divided into three major zones or sub-habitats: littoral, limnetic, and profundal.

(*a*) *Littoral Zone*. The littoral zone adjoins the shore (and is thus the home of rooted plants) and extends down to a point called the light compensation level, or the depth at which the rate of photosynthesis equals the rate of respiration. Within the littoral zone producers are of two main types: rooted or benthic plants, and phytoplankton, or floating green plants, which are mostly algae.

The littoral zone is the home of a greater variety of consumers than are other zones. The zooplankton of the littoral zone is rather characteristic, and differs from that of the limnetic zone in preponderance of heavier, less buoyant crustacea which often cling to plants or rest on the bottom when not actively moving their appendages. Important groups of littoral zooplankton are large, weak-swimming species of Daphnia and Simocephalus, some species of copepods, many families of ostracods and some rotifers.

The nekton of littoral zone is often rich in species and numbers. Adult and larval diving beetles and various adult Hemiptera are conspicuous. Various Diptera larvae and pupae remain suspended in water, often near the surface. Pond fish, frogs, turtles, and water snakes are almost exclusively members of the littoral zone community. Tadpoles of frogs and toads are important primary consumers, feeding on algae or other plant material. Periphyton of the littoral zone exhibit a zonation paralleling that of the rooted plants, but many species occur almost throughout the littoral zone. Among the periphyton forms, for example; ponds snails, damsel fly nymphs and climbing dragonfly nymphs, rotifers, flatworms, bryozoa, hydra, and midge larvae rest on, or are attached to, stems and leaves of the plants.

Another group containing both primary and secondary consumers may be found resting or moving on the bottom or beneath silt or plant debris—for example, sprawling odonata nymphs (which have flattened rather than cylindrical bodies), crayfish, isopods, and certain may fly nymphs. Descending more deeply into the bottom mud are burrowing odonata and ephemeroptera, clams, true worms, snails, chironomids (midges), and other diptera larvae.

(*b*) *Limnetic Zone.* The limnetic zone includes all the waters beyond the littoral zone and down to the light compensation level. The community of the limnetic zone is composed only of plankton, nekton, and sometimes neuston (organisms resting or swimming on the surface).

Phytoplankton producers consist of diatoms, green algae, blue-green algae, and the algae-like green flagellates, chiefly the dinoflagellates. The limnetic zooplankton, consists of few species but the numbers of individuals may be large. Copepods, clodocerans, and rotifers are generally of first importance. But their species are largely different from those found in the littoral zone. The limnetic nekton consists almost entirely of fish. In ponds the fish of the limnetic zone are the same as those of the littoral zone, but in larger bodies of water a few species may be restricted to the limnetic zone.

(*c*) *Profundal Zone.* The bottom and deep water area which is beyond the depth of effective light penetration is called profundal zone. This zone is often absent in ponds.

Since there is no light, the animals of the profundal zone depend on the upper zones for basic food materials. In return, the profundal zone provides rejuvenated nutrients, which are carried by currents and swimming animals to other zones. The major community constituents are bacteria and fungi and three groups of animal consumers:

(a) blood worms, or haemoglobin containing chironomid larvae and annelids,

(b) small clams,

(c) and phantom larvae or chaoborus (corethra).

The first two groups are benthic forms, the last are plankton that regularly move up into the limnetic zone at night and down to the bottom during tbe day.

All animals of the profundal zone are adapted to withstand periods of low oxygen concentration, whereas many bacteria are available.

Lotic Community

The biotic community in streams and rivers is quite different from that in lakes and ponds. The differences in the community are largely due to the differences in physical and chemical conditions of their environments. The presence of definite and continuous current is the main characteristic of lotic habitats. Streams are usually relatively shallow and, therefore, have a large surface compared to their depth. Land-water interchange is relatively more extensive in streams, resulting in a more open ecosystem. This means that streams are more intimately connected with the surrounding land than are most standing bodies of water. For this reason a significant portion of their (stream's) nutrients falls into them from their banks. Another large nutrient source for many streams is the detritus washed in from connected lakes and ponds. Third important difference between streams and ponds is that streams are ordinarily well supplied with oxygen. For this reason stream animal generally have a narrow tolerance and are very sensitive to decreases in the oxygen content of the water.

Current is a major limiting factor in streams. It might be assumed that plankton would be absent from streams, since such organisms are largely at the mercy of the current. Plankton in small streams, if present, originates in lakes, ponds, or backwaters connected with streams and is soon destroyed as it passes through rapids of streams. Only in slow-moving parts of streams and in large rivers plankton is able to multiply and thus become an integral part of the community.

Streams have their own producers, such as fixed filamentous green algae, (Cladophora) encrusted diatoms, and aquatic mosses (Fontinalis). But these are usually insufficient to support the numerous consumers found in streams. Many of the primary consumers in streams are detritus feeders that depend in large part on organic materials which are swept or fall in from terrestrial vegetation. Generally, benthic invertebrates are more numerous in rapids communities, whereas stream nekton and borrowing forms, such as clams, burrowing odonata and ephemeroptera, are more abundant in pools. Stream fishes usually live in pools and feed in or at the base of rapids communities. Most stream living animals are easily recognizable from their adaptations for living in swift current.

CHARACTERISTICS OF FRESHWATER HABITAT

Following are the characteristics of freshwater habitat or the limiting factors of frashwater habitat:

Temperature

In freshwater, the range of temperature variation are smaller and temperature changes occur more slowly as compared to those in air. This is due to the following unique thermal properties of water.

(*i*) *High Specific Heat*, that is a relatively large amount of heat is involved in changing the temperature of water. One gram calories (g cal)˙ of heat is required to raise 1 ml (or 1 g) of water by 1°C (between 15° and 16°C). Only ammonia and a few other substances have values higher than 1.

(*ii*) *High Latent Heat of Fusion.* Eighty calories are required to change 1 g of ice into water and *vice versa.*

(*iii*) *Highest Known Latent Heat of Evaporation.* Five hundred and thirty six calories per gram are absorbed during evaporation of water. The process of evaporation occurs more or less continuously from vegetation, water and ice surfaces. A major portion of the incoming solar radiation is dissipated in evaporation of water from the ecosystems of the world, it is this that moderates climates and makes possible the development of life.

(*iv*) *Water is Densest at 4°C,* it expands and hence becomes lighter both above and below this temperature. This unique property of water prevents deep bodies of freshwaters to freeze solid.

Temperature is one of the major limting factors for aquatic organisms as they are stenothermal (having narrow range of temperature tolerance). The influence of temperature on organisms has been discussed in previous Chapter. Temperature changes produce characteristic patterns of circulation of water and its stratification which greatly influence aquatic life. These will be described later in this chapter. Large bodies of water greatly modify the climate adjacent land areas.

The freshwater habitats occupy a relatively small portion of the earth's surface as compared to marine and terrestrial habitats but their importance to man is far greater due to the following reasons:

1. They are the most convenient and cheapest source of water for domestic and industrial needs (we go more water from the sea but at great cost in terms of energy required and salt pollution created).
2. The freshwater are important in the hydrological cycle.
3. Freshwater ecosystems provide more convenient and cheapest waste disposal system.

Physical factors play an important role which for the most part are responsible for the distribution of animal life in fresh water habitat.

Depth

Obviously, depth has an important bearing on the physico-chemical prospects of water. The penetration of solar rays varies according to the degree of turbidity as well as depth. The photosynthetic regulation and water circulation patterns are very much dependent upon sunlight. On one hand, productivity in shallow ponds is facilitated as light reaches the bottom while on the other, in very shallow ones the same factors inhibit the survival of bio-organic materials.

Water Movements

Water movement plays a vital role in shaping the biocoenosis, because of its effect on the substratum. When the movement is vigorous it will carry away the finer particles leaving only stones. When the movement is weak sand will settle and as it slackens more particles will drop to the bottom. Moss grows on the stones that are not turned over frequently and the higher plants take the root in sand and mud. In the lake, the effect of waves diminishes as the depth increases. The sand region is often sparsely inhabited since the animals are confronted with a difficult substratum with little to eat. Beyond this fragments of dead leaves and other fine particles of organic matter fall to the bottom and provide a supply of food. Animals that live in running water show adaptations of structure and behaviour and inhabit under stones and plants where there is decrease in the water current. They are most numerous at optimum current and are progressively lesser in slower and faster water. Animals that depend on the current to bring food, usually select an optimum speed. Some animals desire a current of certain speed to supply them the oxygen they require. Selection of a substratum of a particular type is common in both still and running water. It may be related to the feeding and reproduction of the animals.

Molar Agents

Molar agents may be in solid or fluid state. Wind, currents, waves, and tides are important molar agents which make the habitats.

Wind

By inducing large scale water movements wind facilitates dispersal of aquatic organisms. Many examples could be ciled.

Indeed, the distribution of immature larval forms is largely due to the churning action of wind on the water bodies. Moreover, the direction of wind is such as to enable the aquatic organisms to obtain their natural food.

Currents

Currents are caused due to winds and rotation of the earth. They help in distribution, selection, and dispersal of vast number of smaller animals. Currents may be seasonal, permanent or adventitious. In fresh water, currents distribute the food and oxygen supply to the organisms. Dissolved oxygen content of rivers, streams and lakes is correlated with the rate of flow of water currents. Stream and pondwater froms react to currents according to the content of oxygen in the water that surrounds them. Shallow lakes with vertical circulation at all the time naturally have different habitat than deep stagnant waters. The rivers having sluggish current show large abundant bottom flora and fauna. Currents above moderate speed limit the plankton population. The fish species in streams are roughly arranged in a graded series—according to the swiftness of the current. The caddis fly larvae are also distributed according to speed of the running water. Sponges develop their shapes according to speed of the currents. In hill districts water current may increase suddenly as a result of heavy rain and decrease again quickly when the rain stops.

Effects on Plankton

Majority of animals that dwell in running water also show adaptive behaviour to life in sheltered places between stones or in vegetation, where there is little current. In rapidly flowing water, where bottom is stony, larvae of simuliidae are morphologically adapted for life in that they have means to cling to a smooth surface and for filter feeding from water to obtain food. Net-spinning tricopterans also make use of the current to obtain their food. *Ancylus*, *Rhithrogena* and *Gammarus* are three commonest species which are found in fairly fast streams. Many fishes are well adapted to life in torrential streams.

Seiches

In standing water wave movements play an important role. They cause larger displacements than the travelling waves. In travelling waves, the water particles move through the same distance but differ in phase. However, in standing waves, phase of all particles is the same but extent of the movements is different. Phenomenon of standing waves occurs in all lakes. This is called seiches. However, only in large water bodies, a periodic rise and fall of the water surface is observed. Such changes are extensive enough to manifest so evidently as can be noted without employment of measuring instrument. Standing waves are not only limited to interface of water and air but also occur within water body mass if different strata having different densities get developed. These are called internal seiches. Internal seiches occur in lakes at interface of epilimnion with lighter water and hypolimnion with heavier water. Phenomenon of seiches is conveniently studied by temperature measurements taken at short intervals of time and hence standing waves are also called as temperature seiches. Importance of internal seiches to limnology lies both in vertical and horizontal movement of water. Seiches depend upon morphology, density, temperature fluctuation of lake water, etc. Oscillations are more rapid in summer stagnation and slacken in autumn circulation and disappear entirely when the density differences get diminished.

Tides

Lunar tides, also called true tides are found in large lakes. Amplitude of lake tides is naturally smaller. Due to the presence of complicated seiches difficulties occur in observing and measuring lake tides. Seiche synchronised with lunar period would enhance the tide range, and conversely when it is out of phase, would inhibit tidal effects. Tides exert influence on river ffow which serves to increase turbulence in the lower stream course and to develop periodic flooding of the valley.

Waves

Currents, wind or combined effect of both cause development of water waves. A series of trichoidal circles produced by surface waves in water become smaller with increase in depth.

Evaporation

Cold air has very little capacity to hold moisture. However, when air becomes warmer, it attains high evaporating power. Winds accelerate the evaporation and drying of water. The rate of evaporation of water depends upon surface of water, deficiency of moisture in the upper atmosphere, temperature, winds, current, etc. The evaporation also depends upon the differences in concentration of water molecules in the interfacial layers of both air and water.

Temperature

Water is a poor conductor of heat. Because of its specific heat, it takes longer time to warm up than the air and slower in cooling so that the temperature fluctuations are neither so great nor so radical in water as in the air. This is of importance to organisms that entirely depend on their surroundings for their body heat. Large variation in temperature is particularly well evident in ponds and brooks, which freeze quickly in severe winter and dry out as fast in the summer heat. In running water habitat there is smaller annual and diurnal temperature variation than is usually encountered in stagnant water. During winter the surface water freezes and the bottom water maintains its live organisms at even 4° C. This ice cap cover protects the lakes from adverse effects of wind. Stratification of temperatures can be seen only in those expansive lakes which suffer large diurnal thermal environmental variations. A factor of the greatest blological importance, in large deep lakes, is the existence in summer, of layers of water at different temperatures. The absorbing power of water prevents the sunlight and heat penetrating into the lower depths, so that two main layers are formed. viz., the upper warmer and lighter waters and colder heavier layer at the bottom. Winds circulate the water to a certain extent but their effect is limited to the upper stratum of the lake and an intermediate layer outside the influence of the wind comes into existence wherein the temperature change is very rapid. This layer is called 'Thermocline', while the water above is called 'Epilimnion' and the colder water below is called the 'Hypolimnion'. Thermocline acts as an effective barrier, preventing the free mixing of the waters, dissolved chemicals and gases between upper and lower layers thus rendering the two layers quite different in their ability to sustain life. The hypolimnion is cold and dark with little plant life. On the other hand, the epilimnion is well lighted, warm and well oxygenated in which multitude of microscopic animals and plants can produce food and multiply. On the death of animals and plants, they fall into the hypolimnion where they decompose by the action of bacteria and use up the oxygen present. As there is no circulation between the two layers, the products of decomposition do not become available to enrich epilimnion. This may sometimes limit the animals the lake can support A lake where such conditions exist is called as Eutrophic lake. On the other hand, where there is scarcity of substances which are needed for the plant growth, a sparse plankton is produced in the upper layers and the decomposition of its remains at the bottom does not use up all the oxygen during the season. Such lakes are called as Oligotrophic lakes. A third type of lake is called Dystrophic lake which is generally a lake with bottom composed of humus ooze and soft acid water, which is brown in colour. Lakes are often seen changing their characteristics due to seasonal fluctuations. The large quantities of dissolved suspended inorganic and organic matter are being washed in the autumn and the winter floods together with the products of decomposition brought into circulation at the overturn and provide conditions favourable for the growth of animals and plants.

Such are the main exciting characteristics of lake biology and the study is not only of absorbing interest but also better understanding contributes to the benefit of mankind.

Turnover

Above 4°C the highest temperature occurs at surface of water while coldest remains at the bottom. This produces a thermal gradient. However, below 4°C the colder water occupies the surface and the thermal gradient in which water gradually becomes warmer increases with depth. Thermal gradients of such type occur not only in winter but also are reversed during summer. A transition between these two gradients occurs in autumn and spring and is known as a 'turnover' or overturn. Strong spring winds stir the water from top to bottom. This phenomenon is called spring overturn.

Summer and Water Stagnation

In sheltered situations in a lake when during long summer periods the upper strata gets warmed rapidly preventing mixing by wind with lower levels and thus stability is quickly established. This is well known as summer stagnation.

During low temperatures the surface layer of ice cover protects the lake from the wind action. Such a permanent stratification in which there is a sharp rise in the temperature of lower strata for long periods is called as winter stagnation. Mixing of water is prevented when the difference in temperature exceeds 10°C.

Other Characters

Light

In natural waters, blue light of spectrum is main illumination which is transmitted in water at the depth of 100 metres. Red and orange colours are transmitted in coloured waters more deeply but their intensity reduces quickly.

TRANSPARENCY OF WATER AND LIGHT PENETRATION

Standing freshwater is often not as transparent as it contains a lot of suspended material which obstructs the passage of light through it. This slows down the photosynthetic activity of green plants which the habitat harbours. The fauna which depends on the flora is also affected. Turbidity, especially when caused by clay and silt particles is often important as a limiting factor. Conversely when turbidity is caused by living organisms, measurements of transparency become indices of productivity.

Shallow standing freshwater habitats *e.g.*, small ponds receive light to the very bottom, this results in an abundant growth of phytoplankton and other rooted plants. The fauna which feed on these plants are also abundant *e.g.*, growth of zooplankton is controlled by the growth of phytoplankton upon which they feed.

Light is the principal factor which controls the diurnal (day and night) movement of plankton although other factors like dissolved O_2, free CO_2, temperature, food, and gravity also have a slight influence. The diurnal movement may be of various kinds. For example, the total plankton population may be in the surface waters in the afternoon and migrate deep downwards at night or it may be movement only upto thermocline or only a few members of the whole population may move, others may not move at all.

Colour and Turbidity

Pure water bodies appear nearly black as they absorb all light components of spectrum. The lake water containing suspended material is seen blue in colour due to the scattering of light by water molecules. However, colour of water ranges from green-blue through blue-green, green with yellow, yellow-brown to brown. Colour of water which appears due to material in colloidal state is known as specific colour. Apparent colour is that which is result of interplay of light on suspended particulate matter together with other factors such as reflection of water base and sky. Further, planktons are responsible for colour of water. For example, presence of abundant blue-green algae gives dark greenish colour to water. Water containing diatoms shows yellowish colour. Zooplankton like micro-crustaceans gives tint of red colour to water. Undecayed organic matter gives brown colour to the water. Colour within any lake may not be uniform from surface to bottom.

Turbidity

Degree of opaqueness developed in water by means of suspended particulate matter is known as turbidity. Substances like humus, silt, organic detritus, colloidal matter, plants and animals brought into lake from outside are called as allochthonous. Substances which are produced within lake and produce turbidity are called as autochthonous. In contrast to lake, river and stream waters show very little variation in colours. However, running water shows very little turbidity which becomes its characteristic feature.

CONCENTRATION OF RESPIRATORY GASES: O_2 AND CO_2

Oxygen is present in water in a dissolved state. It is poorly soluble in water *e.g.*, at 15°C, only about 7 ml of oxygen dissolves in 1000 ml of water.

Oxygen reaches water (1) by direct diffusion or by movements of water like water circulation and wave action, thus moving streams and running rivers have a high percentage of oxygen than the standing waters. (2) Aquatic photosynthetic plants also supply oxygen to water.

Oxygen content of freshwater varies with the season (seasonal variation) and also during the day and night (diurnal variation). Freshwater shows a conspicuous rise in dissolved oxygen when vegetation is dense and sunlight bright. There is a fall in oxygen content in the later part of the day when sunlight fades. In a lake showing winter stratification (to be described later in this chapter) oxygen may be completely absent.

The reduction of oxygen is caused in a number of ways- the major being respiration and decomposition of dead organisms.

In standing pools, there is a lot of decaying vegetation, hence such pools may become completely devoid of oxygen at times. In such cases, oxygen forms a limiting factor for the organisms.

Carbon di-oxide being highly soluble in water is present in a dissolved state as carbonic acid and as carbonates and bicarbonates of various elements like magnesium and calcium. Carbon-di-oxide is formed during respiration of plants and animals and as a result of decomposition of organic matter. Some CO_2, diffuses into water from the atmosphere.

Carbon-di-oxide content of fresh water varies like that of oxygen. Reduction of CO_2 is caused in a number of ways, the major being photosynthesis. A large quantity of CO_2 is fatal to organisms. Molluscs, some insects and some bacteria can precipitate carbonates which accumulate gradually to form marl deposits on the bottom of certain lakes. When these deposits become huge, there is decrease in animal life present in the lake particularly the bottom dwellers.

Buoyancy

Buoyancy depends upon the density and temperature of the water. Water becomes denser and viscous at low temperature and gives greater buoyancy. Aquatic animals are slightly heavier than water and these animals tend to sink in less dense water. Buoyancy of water varies with seasonal changes. Many aquatic animals change their forms according to buoyancy. When water has greater buoyancy, animals have compact bodies, but when the buoyancy is less, animals develop organs which help them in flotation or impede sinking. Protozoans, rotifers and many crustaceans show compact "winter" forms and attenuated or spinose "summer" forms, a phenomenon known as cyclomorphosis which has been dealt with later. Buoyancy plays a vital role in minimising the locomotory adaptations of the aquatic animals. The density of water relieves the aquatic organisms of the necessity of supporting their own weight and enables them to move easily through the water even with such feeble means of locomotion as cilia of protozoans and lashing hairs of flatworms. Many minute forms of both plants and animals spend much of their life floating in the water. The aquatic media are often used as excellent transmitters conducting sounds and vibrations. Some insects such as Corixid bugs and Hygrobia have stridulating organs by means of which they can produce sounds which are believed to be the means of attracting the opposite sex for mating.

Density

The density of distilled water at 0°C, and 760 mm Hg is 775 times greater than air. The density of large water bodies, lakes and rivers varies at different places and also at different times. These differences in the water density are due to variation in temperature and salt contents of the water. There is a liner increase in water density with increase in dissolved substances of salts. The changes in water density due to temperature fluctuation are more important. The density changes more quickly at higher temperatures than at lower ones.

Viscosity

Viscosity of water is one hundred times greater than that of air. Viscosity is cause of the frictional resistance which water offers to the moving animal. The magnitude of viscosity function is proportional to the extent of surface of animal body which is in contact with water, the speed of movement of animal and to temperature.

Surface Tension

The surface tension acts at the water-air interface and develops a biotope. Many plants and animals and their parts are affected by surface tension in many ways, depending upon whether parts of animal and plants are wettable or not. For example, Potamogeton young leaves and Cladocera shell are water repellant, hence water molecules always shy away from shell or leaves, etc., and the surface tension not only prevents wetting but many aquatic animals and plants are able to float and avoid submergence precisely because of this property of their integumental epidermis.

OTHER DISSOLVED CASES

Other dissolved gases are methane (marsh gas), hydrogen sulphide, ammonia and nitrogen.

Methane is formed due to decomposition of organic matter at the bottom and is seen coming out of the surface as bubbles. It is not toxic to the organisms in small amounts.

Hydrogen sulphide is formed due to decomposition in the absence of oxygen *i.e.*, anaerobic decomposition of organic matter containing sulphur. It characterises stagnant fresh water pools where oxygen is depleted or in streams and lakes contaminated with sewage. This gas is highly toxic to organisms.

Ammonia is formed due to the decomposition of organic matter or due to industrial waste being thrown into water. It is toxic in high concentrations.

Nitrogen gets diffused into water from the atmosphere. It is not toxic to organisms in general.

Concentration of Biogenic Salts

Salts are present in water in dissolved state. Except for certain mineral springs, even the hardest fresh water have a salt content of less than 0.5 part per thousand, compared with 30 to 37 parts per thousand for sea water. The important salts are compounds of nitrogen, phosphorus, calcium and other minerals present in traces.

Salts reach water by erosion, inflow and by decay of aquatic forms. From this, it follows that salinity will vary in different fresh waters depending upon their place of occurrence.

Nitrogen salts occur as nitrates and nitrites. The growth of phytoplanktonic organisms particularly that of blue-green algae depends on nitrogen content of water. Excess of ammonium salts in water is fatal to organisms. Thus nitrogen may be a limiting factor.

Phosphorus forms a limiting factor for organisms. For example if phytoplankton become abundant they utilise most of the phosphorus and this results in a total elimination of other plants which require it.

Calcium occurs as carbonates and bicarbonates. In soft water calcium content is about 9 mg/litre, in hard water it may reach 26 mg/litre. Molluscs need a lot of calcium for shell formation, thus they are abundant in hard water than in soft water. Calcium plays an important role as it is essential for metabolic activity of organisms and controls the pH of aquatic environment.

Other minerals present in water which are required by organisms in traces are magnesium, manganese, iron, sodium, potassium, sulphur and zinc. These minerals are important for the physiological activities of the organisms.

Salinity and Osmoregulation: The concentration of salts in fresh water is lower than that of body fluids of animals *i.e.*, fresh water is a hypotonic medium. On account of this difference in salt content water tends to enter the body of organisms by osmosis. Fresh water organisms have therefore developed special mechanisms to eliminate excess of water. For example fresh water protozoa have contractile vacuoles to get rid of excess of water, in fresh water fishes kidneys serve this purpose of osmoregulation. The marine animals have not been able to invade fresh water due to difficulties in osmoregulation. A few exceptions are the bony fishes, some birds and mammals who have reinvaded the sea. Their body fluids have a salt content lower than that of sea water (=hypotonic) but they have evolved a metabolic osmoregulation mechanism which involves excretion of salt and retention of water.

pH

The pH of freshwater is influenced by soluble organic and inorganic substances and forms a limiting factor for organisms.

Wind and Currents

The wind influences the standing freshwater habitats as it causes water currents. This effect of wind depends upon the extent of exposed water surface. presence or absence of an upland protecting the standing waters and its configuration. Water current determines the distribution of phytoplankton and zooplankton, salts and gases throughout the body of water.

CLASSIFICATION OF LAKES

(*a*) Classification based on water circulation patterns.

(*b*) Classification based on primary productivity.

(*c*) Special types of lakes.

(*a*) *Based on water circulation patterns*, the lakes of the world can be divided into the following categories:

(*i*) *Dimictic lakes* (mictic = mixed): Two seasonal periods of general circulation (=overturns).

(*ii*) *Cold monomictic lakes*: Water never above 4°C as in polar regions. Seasonal overturn in summer.

(*iii*) *Warm monomictic*: Water never below 4°C as in warm temperate or subtropical lakes. One period of circulation in winter only.

(*iv*) *Polymictic lakes*: More or less continually circulating with only short (if any) stagnation periods as in high altitude equatorial lakes.

(*v*) *Oligomictic lakes*: Rarely or very slowly mixed *i.e.*, thermally stable *e.g.*, many tropical lakes.

(*vi*) *Meromictic lakes*: Permanently stratified, mostly due to chemical difference in hypoliminion and epilimnion waters. The chemical difference is due to salts liberated from sediments which reach water and creat a permanent density difference between surface and bottom waters. In this case boundary between circulating and non-circulating layers is the chemocline instead of thermocline. Examples are Big Soda Lake in Nevada and Hemmelsdorfersee lake in Germany.

(*b*) *Based on primary productivity* which depends on regional drainage, geologic age and depth, lakes may be 'Oligotrophic' (=few foods) and 'Eutrophic' (=good food) lakes. These lakes differ as follows (Table on the next page).

(*c*) *Special types of lakes*: The special types of lakes may be classified into the following seven types:

(*i*) *Dystrophic lakes*: These include brown water, humic and bog lakes. They have a high concentration of humic acid in water. Bog lakes have peat filled margins (where pH is usually low) and develop into peat.

(*ii*) *Deep Ancient lakes*: Lake Baikal in Russia is the deepest lake in the world, it was formed by earth movement during mesozoic era. This lake has many endemic (found nowhere else) species.

(*iii*) *Desert salt lakes*: In these lakes due to lot of evaporation, salt concentration is high. Examples are Great Salt Lake in Utah. It has Brine shrimp (Artemia) which tolerates high salinity.

(*iv*) *Desert alkali lakes*: These occur in igneous drainages in dry climates. They have high pH and Carbonates *e.g.*, pyramid lake in Nevada.

(*v*) *Volcanic lakes* are acid or alkaline lakes associated with active volcanic regions. They have extreme chemical conditions and restricted biota *e.g.*, some Japanese and Philippines lakes.

(*vi*) *Chemically stratified lakes* (=meromictic or partly mixed lakes) have been already described.

(*vii*) *Polar lakes* have surface temperature below 4°C or temperature rises above it only during brief summer when ice melts and water circulation occurs. Plankton population grows during this, they store fat for long winters.

Classification of Lakes Based on Thermal Cycles

The lakes are classified into following types depending upon thermal cycles.

Amictic lakes—are covered over by ice and hence get protected from weather and other external physical factors. Usually such lakes are found at high altitudes and in temperate zones.

Cold monomictic lakes—are found in polar regions and have temperature 4°C at any depth. They exhibit inverse temperature stratification in winter. Even when stratification occurs, temperature never goes beyond 4°C in summer also.

Warm monomictic lakes—are formed in warmer latitudes. Water temperature, at any depth never falls below 4°C. Only one circulation per year takes place.

Dimictic lakes—occur in temperate zone and in higher altitudes in subtropical regions. In these lakes spring and autumn circulation occurs every year. Thermal stratification is inverse in winter and direct in summer. The temperate oscillates on either side of 4°C during the year.

Oligomictic lakes—are common in tropical zones. Temperature remains always higher than 4°C. Circulation occurs rather irregularly.

Polymictic lakes—are located in high mountains of equatorial regions. The temperature remains a little over the 4°C and mixing goes on continuously. Uniform stratification does not occur in such type of lake.

Meromictic lakes—Some lakes have sharper concentration between epic- and hypolimnetic zones and in sufficiently deep basins are set up stabilities which prevent fall turnover. Circulation occurs only to a limited depth. Such lakes are called meromictic lakes.

CLASSIFICATION OF ANIMALS

Neustonic Animals

Surface of the water bodies, as an interface between air and water is responsible for harbouring an interesting and diverse community of plants and animals. The organisms which occur on this surface film of water are called neustonic organisms and the ecological area as neuston. The animals and plants which live under surface of water film are called as infraneustonic. Infraneuston often includes such organisms as snails and flatworms. Seasonally, it may also include larvae and pupae of Dipterous insects, species of hydra, Entomostraca and variety of organisms. There is free floating plant community including plants like Lemna and Azolla. Number of insects coexist with these plants. There also occur swimming whirligig beetle. Such organisms occurring on the upper surface of water film are called Supraneustonic. The large animals in supraneuston are water-skaters, water-striders, Hebrids, Springtails, certain spiders, amphibian tadpodes and Aplocheilus type of fishes. Certain of these animals have power to break the surface of film water and swim below for a short period. The minute supraneuston include bacteria, fungi and certain algae.

Nektonic Animals

Organisms of large size which are capable of swimming freely independent of water currents and determine their own distribution regardless of water movements are called as nektonic organisms or nekton. Most of nekton of inland water is composed of fishes, amphibian larvae, reptiles, etc. In limnetic nekton, animals occupy almost whole of the open water. Few reptiles like snakes and turtles are also observed, swimming in water but they generally confine to the shores. In southern regions alligators and crocodiles occupy fresh water bodies. These nekton is more widespread and in larger numbers in littoral zones of water. Greatest diversity of species in nekton is observed here in littoral zone. In addition to fishes, number of free invertebrates including insects are present in this zone. In littoral zone, where "pond weed" vegetation is abundant and contains largest nekton population, certain birds like gulls, ducks and others swim temporarily in water. If the term nekton includes temporarily swimming animals, these birds should be included in nekton. Distribution of nekton in different layers of water bodies depends upon season, physiological state of animals and stage in their life history. Certain fishes occupy surface region, others intermediate regions and still others are found in deeper and coldest water. Fishes such as lake trout and catfishes occur in the deeper water of great lakes.

The Aufwuchs (=Periphyton)

The organisms, which are attached to or creep upon a submerged substrate are called as aufwuchs. Benthos is not included in this aufwuchs community. Ecologically, base of aufwuchs is made up of unicellular and filamentous algae. On this base in between plants various protozoans, bryozoans and rotifers thrive. Free living organisms like roundworms, crustaceans, annelids are also found in the aufwuchs. The number and kinds of animals vary with substrate, water movements, depth and also chemical properties of waters. For example, bryozoans generally inhabit the substrate formed of a sulphur containing algae like Chara. The Pelmatohydra are found in aufwuchs having substrate of dense vegetation including Vallisneria and algae. Thus, kinds of animals depend upon the nature of substrates. Aufwuchs on stones in brooks of moderate current and aufwuchs on leaf as substrate show corresponding differences in animal composition. On stones aufwuchs show dense growth of algae and diatoms. Animals which inhabit such surface are mayflies, nymphs and blackfly larvae, immature odonata as also snails and flatworms.

Aufwuchs on leaves show dense association of filamentous organisms such as stalked protozoans and hydra. In lakes and ponds aufwuchs found on submerged plants or their parts are very rich and variable. Water which contains high concentration of bicarbonates shows absence of certain sessile rotifers. Dissolved oxygen, carbon dioxide and other chemicals are limiting factors for distribution of periphyton animals. This observation provides an illustration of how chemical constituents affect the animal life.

ECOLOGICAL CLASSIFICATION OF FRESHWATER ORGANISMS

1. *Based on their position in the food chain,* the freshwater organisms are classified as follows:

(*i*) *Autotrophs* (*Producers*): Green plants and chemosynthetic microorganisms.

(*ii*) *Phagotrophs* (*Macroconsumers*): Primary, secondary etc, These include herbivores, predators, parasites etc.

(*iii*) *Saprotrophs* (*Microconsumers*) or (*Decomposers*): These are subclassified according to the nature of the organic substrate decomposed.

2. *Based on the mode of life,* the fresh water organisms may be classified as follows:

(*i*) *Benthos*: It includes organisms which are attached to the bottom or rest on the bottom or live in the bottom sediments. The benthic animals may be subdivided according to the mode of feeding, into filter feeders (*e.g.* clam) and deposit feeders (*e.g.* snail).

(*ii*) (*Periphyton*): Organisms (both plants and animals) which are attached to or cling to stems and leaves of rooted plants or other surfaces projecting above the bottom.

(*iii*) *Plankton*: Floating organisms whose movements are more or less dependent on currents. Depending upon the size of organisms, plankton may be of the following 2 types:

(a) Net Plankton (Macroplankton and microplankton) which is caught in a plankton net (fine meshed net) when it is towed slowly through water.

(b) *Nannoplankton* which is too small to be caught in plankton net, it is therefore collected in a bottle or by means of a pump.

(*iv*) *Nekton*: Swimming organisms able to swim at will, hence capable of avoiding plankton nets, water bottles etc. Examples are fish, amphibians, large swimming insects and so forth.

(*v*) *Neuston*: Organisms resting or swimming on the surface.

ORGANISMS OF LENTIC HABITAT

1. Biota of Littoral Zone

(*i*) *Producers*: The producers of littoral zone are of two main types:

(a) Rooted benthic plants which are mostly spermatophytes (Dn. Spermatophyta)

(b) Non-rooted plants which comprise the phytoplankton (floating green plants) and filamentous green algae (pond scum)

(*ii*) *Consumers*: (a) Periphyton, (b) Zooplankton, (c) Nekton, (d) Neuston.

(*i*) *Producers*

(*a*) *Rooted benthic plants* typically form concentric zones within the littoral zone. A representative arrangement proceeding from shallow to deeper water is given below (Fig. 3.1) but it should not be assumed that all three zones will be present or they will be arranged in the order given here.

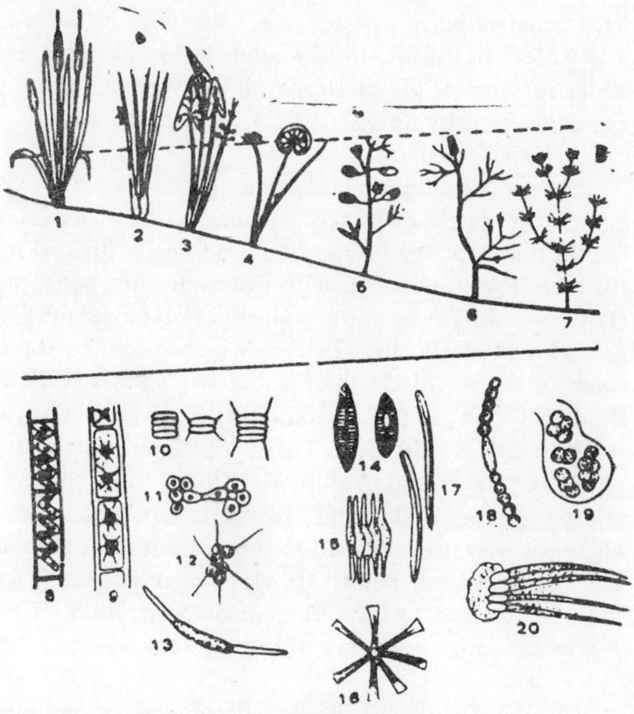

FIG. 3.1: Producers of Littoral Zone of Ponds and Lakes.

Zone of Emergent Vegetation: Consists of rooted plants with principal photosynthetic surfaces projecting above water. They obtain CO_2 for photosynthesis from air and raw materials from beneath the water surface *e.g.* cat tails (genus Typha), bulrush (scirpus) arrowheads (Sagittaria) (1, 2, 3, respectively in Fig. 3.1) and burreeds (spargamum), spike rushes (Eleocharis) and Pickerel weeds (Pontederia). The emergent plants together with those on the moist shore form an important link between water and land environments. They are used for food and shelter by amphibious animals and provide a convenient means of entry into and exit from water for aquatic insects which spend part of their lives in water and part on land.

Zone of Rooted Plants with Floating Leaves. Ecologically this zone is similar to the zone of emergent vegetation but here the leaves float horizontally on the surface and this reduces light penetration into water *e.g.* water lily Nymphea (4 in Fig. 3.1) which has four species and water shield Brasenia. The under surfaces of lily pads provide convenient resting places for egg deposition by animals.

Zone of Submergent Vegetation comprises rooted or fixed plants completely or largely submerged. The leaves of vegetation in this zone tend to be thin and finely divided or adapted for exchange of nutrients with water. In this zone, pond weeds (Potamogetonaceae) are prominent. Potamogeton is the largest genus having sixty five species which occur in the temperate zones of the world. In the Fig. 3.1 two species of Potamogeton (P. diversifolia and P. pectinatus) have been shown (5 and 6 respectively in Fig. 3.1). Other common weeds are Ceratophyllum, Najas and Vallisneria Chara (muskgrass) and the related genera Nitella and Tolypella are generally classes as algae yet they are attached to the bottom and their life resembles that of higher plants. In view of this chara (7 in Fig. 3.1) may be classed with submergent vegetation. Chara often marks the inner boundary of littoral zone.

(*b*) *The Non-rooted Plants*: *Phytoplankton*, comprise many species of algae found floating throughout the littoral zone. Some species of algae are found attached to or associated with rooted plants, these are especially characteristic of littoral zone. The chief types of algae are the following:

Diatoms (Bacillariaceae) with yellow or brown pigment in the chromatophores which are so abundant that they mask the green colour of chlorophyll. These are good indicators of water quality *e.g.*, Navicula, Fragilaria, Asterionella (which floats in water like a parachute) and Nitzschia (14, 15, 16, 17 respectively in Fig. 3.1).

Green algae (Chlorophyta) having both floating and attached forms. In these, the chlorophyll is not masked by other pigments, therefore populations of green algae have a bright green appearance *e.g.* Scenedesmus, Coelastrum, Richteriella and Closteria (a desmid) (10, 11, 12 and 13 in Fig. 3.1).

Blue green algae (Cyanophyta) having diffused chlorophyll in the algal cells (and not concentrated into chromatoplasts) which is masked by blue green pigment. In polluted ponds and lakes this may develop into blooms. These algae are able to fix up gaseous nitrogen into nitrates (on land the bacteria of soil fix up atmospheric nitrogen into nitrates). Many of these blue-green algae are not palatable, hence not eaten by consumers. This is one of the reasons for the formation of large blooms. The cells of these living algae excrete metabolities which are toxic and give a bad taste and odour to drinking water. After death, during decay the breakdown products of these algae are also toxic and give a bad taste to water. The common genera are Anabaena, microcysts, Gloeotricha (18, 19, 20 respectively in Fig. 3.1), Oscillatoria and Rivularia may cover the bottom or may be found attached to stems and leaves of submerged spermatophytes (seed plants).

A sample of free floating phytoplankton of littoral zone reveals numerous diatoms, desmids and other green algae, chlorophyll containing protozoa like Euglena and its many relatives.

Filamentous Green Algae or Pond Scum comprise the common genera namely spirogyra, zygonaema (8 & 9 respectively in Fig. 3.1) and Oedogonium. Other filamentous algae are found attached to rooted plants like Chara, this association is a commensal or mutualistic relationship. Such algae assume a periphyton life. When a pond or a lake is polluted with excess of nutrients, the filamentous green algae often develop huge blooms, that rise to the surface as they are buoyed up by the entrapped oxygen. They are now said to form pond scum. Now the oxygen produced by photosynthesis largely escapes to the atmosphere. When the bloom dies oxygen in water is used up, this results in mass killing of fish. Blooms are thus very harmful to pond life.

(*ii*) *Consumers*

Periphyton exhibit a vertical zonation in contrast to the horizontal zonation of rooted plants although many species occur throughout the littoral zone.

One group of periphyten comprises both primary and secondary consumers like pond snails, damsel fly nymphs, climbing dragon fly nymphs, rotifers, flat worms, bryozoa, hydra, midge larva which rest on or attached to leaves of large plants (Fig. 3.2).

The second group of periphyten also comprises both primary and secondary consumers which rest or move on the bottom or below silt or plant debris *e.g.* odonata nymphs, crayfish, isopods and certain May fly nymphs.

The third group of periphyten is met with on descending down more deeply into the bottom mud. These include burrowing odonata and ephemeroptera, clams, annelids, snails and especially chironomids (midges) and other Diptera larvae which live in minute burrows.

Zooplankton of littoral zone is characteristic and differs from that of limnetic zone in that the invertebrate groups of zooplankton are large, weak swimming species of water fleas (cladocera) like Daphnia and Simocephalus, copepods, ostracods and some rotifiers.

Nekton of littoral zone is rich in species numbers. It comprises of insects and vertebrates. The conspicuous insects are adult and larval diving beetles and various adult Hemiptera, some of these are carnivorous and others are partly herbivorous or scavengers. Larvae and pupae of various diptera remain suspended in water often near the surface. The vertebrates are mostly amphibious like frogs, salamanders, turtles and water snakes. These cold blooded vertebrates become more

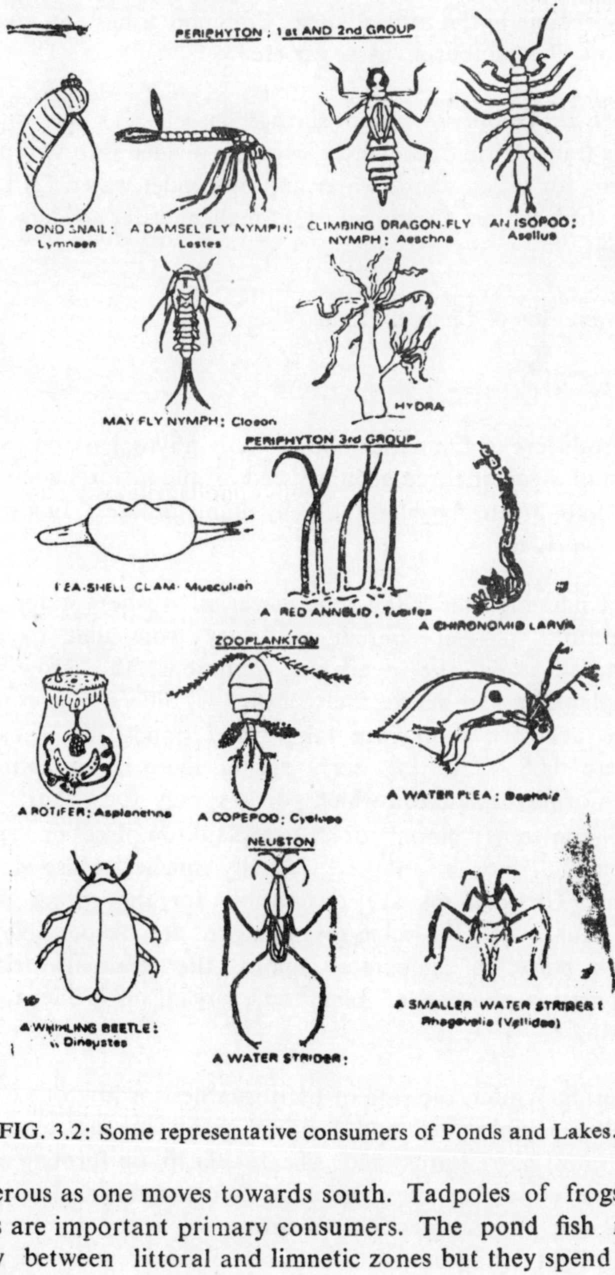

FIG. 3.2: Some representative consumers of Ponds and Lakes.

numerous as one moves towards south. Tadpoles of frogs and toads are important primary consumers. The pond fish move freely between littoral and limnetic zones but they spend most of their time in the littoral zone. Common fishes are sunfish, minnows like gambusia, bass, gar etc.

Neusten comprises three surface insects (1) the whirling beetles (fam Girinidae) whose eyes are divided into two parts, one part for vision above water and one under water. (2) Large water striders (fam Gerridae). (3) Smaller water striders (fam veliidae).

(2) Organisms of Limnetic Zone

(*i*) *Producers*

The producers of Limnetic zone are only phytoplankton which consist of algae of three groups listed in the littoral zone and algae like green flagellates chiefly dinoflagellates, Euglenidae and volvocidae.

The limnetic zone is the zone of variation where water level, temperature, oxygen content etc. vary from time to time. Depending upon the available conditions, the growth of phytoplankton and hence the consumers differs. For example in the northern temperate lakes and ponds like those in northern U.S.A., during early spring there is an abundant growth of phytoplankton which persists only for a short time. This is known as 'bloom' or 'phytoplankton pulse' or 'spring flowering'. There is another, usually smaller pulse in the autumn. The diatoms are responsible for the spring bloom while Anabaena: the blue-green algae are responsible for autumn pulse. In temperate regions, the seasonal variation in population density of planktonic organisms is due to the following:

During Winter, the rate of photosynthesis is low due to low water temperature and reduced light but nutrients like nitrates, phosphorus and silicates keep on forming continuously due to the action of microorganisms thus they increase in concentration. In spring when the temperature and light conditions are favourable, plankton show a rapid sudden increase since there is an abundance of nutrients. Very, soon, the nutrients are exhausted so the bloom disappears. Again nutrients accumulate and another bloom is seen in the autumn when conditions are favourable.

Turbulence or upward current movements of water caused by temperature differences in limnetic zone and in keeping phytoplankton near the surface where maximum photosynthesis can occur.

(*ii*) *Consumers*

Zooplankton consists of only few species but the number of individuals may be large. Copepods, cladocerans and rotifers rank first in order of abundance, then come the crustacea but the species are different from those found in the littoral zone. For example the copepod Diaptomus and Cyclops are common genera, the cladoceran Diaphanosoma, Sida and Boomina are common. Mott of the crustacea are 'strainers', they filter bacteria, netritus and phytoplankton by means of combs of setae on their thoracic appendages. The zooplankton are predators, they 'graze' the phytoplankton in the same manner as cattle do on land.

Zooplankton also exhibit bloom along with phytoplankton or immediately after phytoplankton blooms as they largely depend on them. Vertical diel migration is a characteristic feature of limnetic zooplankton.

Nekton consists almost entirely of fish. In ponds, the limnetic fishes are the same as littoral fishes but in larger bodies of fresh water a few species may be characteristic of limnetic zone. Most adult freshwater fishes feed not on plankton but on animals of good size, only a few like Dorosoma and Signalosa are plankton feeders.

(3) Organisms of Profundal Zone

All the animals of profundal zone are adapted to wi hstand periods of low oxygen concentration and many bacteria are anaerobes *i.e.*, able to carry on without oxygen.

(*i*) *Producers*

The producers are absent in the profundal zone as there is no light.

(*ii*) *Consumers*

Consumers of profundal zone depend on littoral and limnetic zones for food. In return, the profundal zone provides 'rejuvenated' nutrients which are carried by current and swimming animals to other zones. The consumers belong to three groups:

(a) Blood worms or haemoglobin containing chironomid larvae and annelids.

(b) Small clams (fam, sphaeriidae).

(c) Phantom larvae (chaoborus): Plankton forms which have a float of four air sacs. The air in sacs also provides a reserve supply of oxygen. These larvae move up into the limnetic zone during night and down to the bottom during day. The adult of these larvae are land dwelling midges (Diptera).

(*iii*) *Decomposers*

Decomposers comprise bacteria and fungi. These are especially abundant in the water-mud interphase where organic matter of littoral and limnetic zones accumulate. They break down the organic remains into simple inorganic substances which are made available to the producers again.

ORGANISMS OF SWAMPS

The edge of lake or pond consists of swampy (water logged) ground outer to which is moist soil. Not much of literature is available on life in swamps. The following is a brief account.

Producers of the swampy ground are plants like arrowheads, cat tails (Typha) and spikerush (Eleocha) and tall grasses known as reeds. Rice plants and plants like polygonum and scirpus grow very well in marshy areas. This plant is in fact a marshy grass and rice fields are merely artificial ecosystems.

The moist soil bears water grasses and plants like Ranunculus, polygonum and scirpus.

Consumers of swamps consist of protozoans, nematodes, crustaceans, insects, snails, live fishes (=air breathing fishes which have accessory respiratory organs for breathing and storing air) and turtles. Swamps are often visited by amphibians and ducks.

Animals living in swamps have adaptations to tide over periods of draught. For this, some have resistant eggs and cysts, some like the African lung fishes survive draught by burrowing in mud and getting surrounded by a slimy muddy cocoon. Water pollution is another hazard in swamps, under such conditions only autotrophic ciliates and holophytic flagellates survive. Marshy areas seen in village near human habitations, contain drainage water, containing disease germs and provide breeding places for mosquitoes.

THE POND

Types of Ponds

Depending upon their origin, ponds are of the following three types:

1. *Ponds Derived from Large Lakes or Streams.* These ponds are continually being formed *e.g.*, when a stream shifts its position, it leaves the former bed isolated. These ponds are quite productive and attract a number of fishermen.

2. *Small ponds* which are not connected to any body of water right from the time they originated *e.g.*, natural ponds which develop in limestone regions where depressions develop in the rocks.

3. *Artificial ponds* are those which are created by man by damming of a stream or basin or by animals like beaver. Artificial ponds are rarely deep enough for stratification. These ponds are of no interest to ecologists as the variety of organisms is reduced to favour large number of desired species particularly fish. The beaver ponds were abundant in America when the European man arrived.

Depending upon their duration, ponds are of two types:

1. *Temporary ponds* are those ponds which remain dry for a part of the year. These ponds support a unique community. The organisms of such ponds reproduce in a short time when water is available and tide over the dry period either by moving out and reaching another pond having water or by lying in a dormant state in the dry soil and becoming active when water is available *e.g.* Fairy shrimp (Eubranchiopoda). When a temporary pond exists only during spring, it is called a Vernal pond, when it contains little water throughout the year but freezes during winter, it is called aestivial pond.

2. *Permanent ponds* are those which contain water throughout the year.

THE LAKE

Lakes may be natural or artificial (=man made). The artificial lakes have turbid water and fluctuating water level. Benthos is lesser than those of natural lakes. Primary productivity and fish yield differ in different artificial lakes. Following is an account of 'stratification' of natural lakes (a feature which is so characteristic particularly of temperate lakes) and classification of lakes.

STRATIFICATION IN TEMPERATE AND TROPICAL LAKES

(1) Stratification in Temperate Lakes

In temperate lakes, temperature variations are pronounced, hence they show summer and winter stratifications as follows:

During Summer, the surface waters become warmer than bottom waters hence only warm top water (epilimnion or surface lake) circulates. This top circulating water does not mix with the more denser colder bottom water (=hypolimnion or under lake). This state of affairs results in the creation of a zone with a deep temperature gradient (=Thermocline) between upper warm and lower cold water (Fig. 3.3).

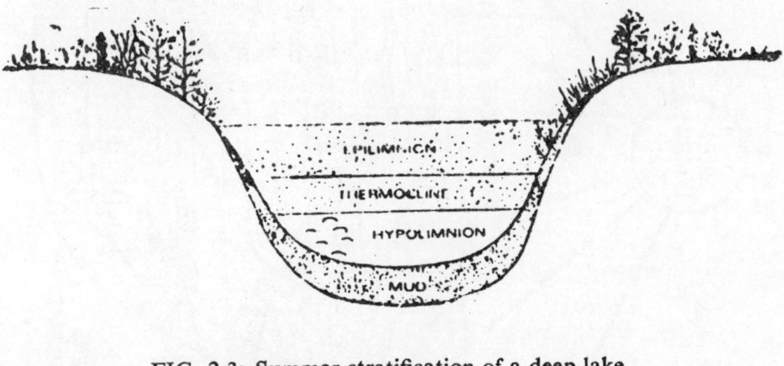

FIG. 2.3: Summer stratification of a deep lake.

During hot months of summer, there is a strong drop in temperature at the thermocline (at least 1° per metre depth of water). Asthe temperature of epilimnion kises further, the difference of temperature between epilimnion and hypolimnion also increases. Since Thermocline acts as a barrier for the exchange of materials between epilimnion and hypolimnion, most of the organisms migrate into the hypolimnion during the summer months. As the thermocline is usually below the compensation level (=range of effective light penetration), the green plants and

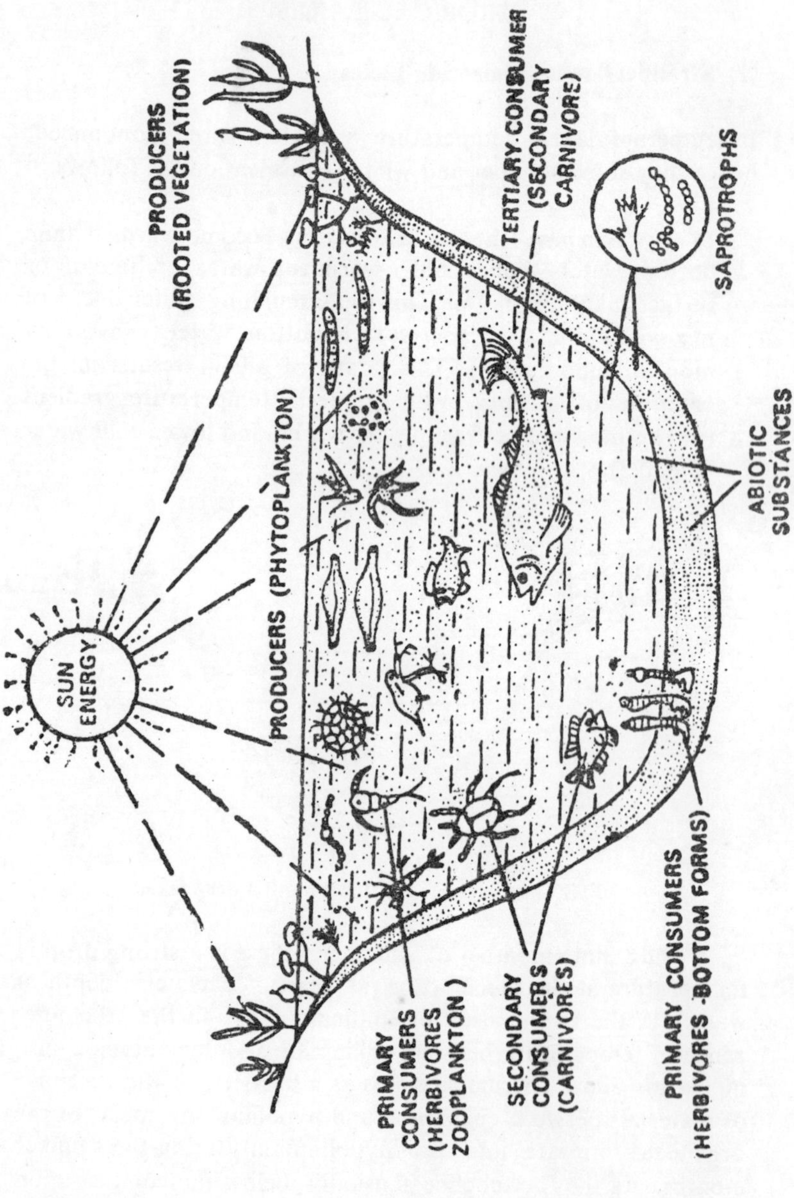

FIG. 3.4: The pond ecosystem.

so the oxygen supply are cut off in the region of hypolimnion. This is referred to as the period of summer stagnation in the hypolimnion. If the water of a lake is very transparent and the thermocline is above the compensation level, light reaches the upper part of hypolimnion, the phytoplankton now grows here and oxygen may be present in greater abundance than on the surface (cold water holds more oxygen). It is to be noted that the euphotic zone does not necessarily coincide with the epilimnion. Euphotic zone *i.e.*, producer zone is based on light penetration and epilimnion on temperature. These zones however coincide roughly during summer stratification of lake.

The extent of depletion of oxygen in the hypolimnion depends on the amount of decaying matter and the depth of thermocline. The lakes which are highly productive (*e.g.*, Great Lakes of U.S.A.) are subject to greater oxygen depletion than the 'poor' lakes as in the former the 'rain' of organic matter from the limnetic and littoral zones into the profundal zone is greater. In productive lakes, cultural eutrophication, hastens oxygen depletion in profundal zones. In poor lakes the steno-thermal fishes (low temperature tolerant) can survive in the hypolimnion due to the presence of oxygen. In productive lakes, these are the first animals to go on eutrophication. This is referred to as the 'Summer kill of the fish'. In contrast to fish, the lower animals of profundal zone are adapted to withstand oxygen deficiency for long periods.

With the onset of winter, temperature of epilimnion drops until it is the same as that of hypolimnion. Now water of the entire lake begins to circulate and oxygen supply is returned to the hypolimnion. This is the 'fall overturn' as now the animals feed and reproduce. During colder months of winter, surface water cools below 4°C, it expands, becomes lighter so remains on the surface and finally freezes (water is densest at 4°C). The bottom water is still at 4°C, thus providing a suitable habitat, for stenothermic animals. This is winter stratification.

During winter stratification, photosynthetic rate is slowed down as the producers do not receive enough sunlight due to a

Difference Between Oligotrophic and Eutrophic Lakes

S. No.	*Oligotrophic Lakes*	*Eutrophic Lakes*
1.	These lakes have changed very little since the time of their formation *i.e.*, they are 'geologically young'.	They are geologically old as they have changed a lot since their formation.
2.	Deeper	Shallower
3.	Hypolimnion is larger than epilimnion.	Hypolimnion is shallower than epilimnion.
4.	Low primary productivity.	High primary productivity.
5.	Littoral vegetation scarce.	Littoral vegetation is more abundant.
6.	Plankton population density is low although number of species may be large.	Plankton population is denser.
7.	Plankton blooms are rare as nutrients rarely accumulate.	Plankton blooms are characteristic as nutrients accumulate in large quantities.
8.	Hyploimnion is not subject to severe oxygen depletion hence stenothermal-cold-water bottom fishes like trout are restricted to the hypolimnion of these lakes.	Hypolimnion is subject to severe oxygen depletion so much so that summer stagnation may exclude cold water-fishes.
	Examples: Great lakes and Finger lakes of New York.	Examples: Lake Mendota and Linsley pond.

layer of ice at the top. Oxygen supply is however not greatly reduced as water holds more oxygen and respiration of organisms and bacterial decomposition are slow at low temperatures. This is referred to as the 'period of winter stagnation' in the hypolimnion. Winter stagnation is generally not so severe as summer stagnation. Sometimes, when snow covers the ice at the top, light is cut off and photosynthesis stops completely resulting in oxygen depletion and 'winter kill of the fish'.

Differences Between Ponds and Lakes

S. No.	*Pond*	*Lake*
1.	Pond is a reservoir of rain water found in many regions with adequate rainfall.	Lake is a natural source of water among hills.
2.	Pond is much smaller than a lake.	Lake is large in expanse.
3.	It is shallow.	It is deeper than a pond.
4.	Littoral zone is large while limnetic and profundal zones are either small or absent.	Limnetic and profundal zones are larger in comparison to littoral zone.
5.	Thermal stratification is either very little as in deep ponds or absent.	Lakes show thermal stratification during summer and winter.
6.	Exposed water surface is much lesser as compared to that of a lake hence wind does not cause water current *i.e.*, pond waters are quiet.	Exposed water surface is large hence wind causes water currents.
7.	Ponds may be temporary or permanent.	Lakes are always permanent.

With the onset of spring, ice melts, water becomes warm, when temperature reaches 4°C, it being heavier sinks to the bottom and water of the entire lake begins to circulate once again.

This is the 'Spring overturn' as now the animals feed and reproduce. In America and Eurasia, two overturns in a year are seen as described above but not in all temperate lakes.

2. Stratification in Tropical Lakes

In sub-tropical lakes, the surface temperature never falls below 4°C. These lakes show a distinct thermal gradient from top to bottom. There is only one general circulation period. In deep tropical lakes the surface temperature is high (20 to 30°C). These lakes show a weak thermal gradient from to to bottom, this results in water density differences which results in a stable stratification all through the year. General circulation is irregular, it occurs mostly in cooler seasons. The very deep tropical lakes tend to remain only partly mixed *i.e.*, permanently stratified.

LENTIC HABITAT VERSUS LOTIC HABITAT

1. The characteristic feature of running water is the movement of water in only one direction. The velocity of water current varies with depth. Normally the maximum velocity is observed in the first 1/3rd depth of a river. The velocity becomes gradually reduced as depth increases and after a certain depth the velocity is almost zero *i.e.*, standing water conditions result. The water current velocity may be uniform throughout or may vary in different parts of the same water and from one time to another. For example, a river may have a lazy flow at a place and a huge water fall at the other where the slope of the ground is very steep. The current velocity may differ at two different faces of a stone which is submerged in water of a river or even in the interspaces of stones.

Current is the chief factor which accounts for the big difference between lentic and lotic habitats, it is also responsible for the differences in the different parts of a lotic habitat. At rocky or sandy shores of lakes especially when rooted plants are absent, velocity of water may be even double that of stream. In such cases therefore the organims which are considered as pond organisms might live in stream and *vice versa.*

2. The thermal and chemical stratification is very little or absent in lotic habitats.

3. Although green plants are not abundant oxygenation is rich and almost uniform throughout due to the large surface exposed to air and constant motion of water. This aids in easy respiration, it is due to this that the animals of running water are susceptible to reduced oxygen content and cannot live in standing water or organically polluted part of stream where oxygen supply is reduced.

4. The depth of water and cross-section area of streams is much less than that of lakes, the land water junction is relatively larger in proportion to the size of a stream. This means that a stream is more intimately associated with the surrounding land than are most standing water bodies. Many of the streams depend on land areas and their ponds, back-waters and lakes for the supply of phytoplankton. From streams energy goes to land and also to lentic habitats in the form of emerging insects and stream dwelling animals removed by the predators. Thus the streams are intimately associated with terrestrial and lentic habitats.

5. A running water exhibits two major zones: the rapids and the pool.

ORGANISMS OF LOTIC HABITAT

As mentioned earlier, the lotic habitat has two zones: the rapids and the pool. The species composition of communities in the rapids differs totally from those of ponds, lakes and also of pools. Since pool communities contain some organisms which also occur in the ponds, the organisms of rapids are said to be typical stream organisms. The organisms (and their density) of rapids and pools are different due to the different types of bottom available whether sand, silt, pebbles, clay, rubble rocks (loose stones etc. as covering of some rocks) etc. For example if the bottom is of sand and silt (least favourable bottom), the number of species and that of individuals (plants and animals) is very small. If the bottom is of rubble rocks the number of species and density of bottom dwelling organisms is the highest. If the bottom is soft as in pool areas, only benthic burrowing form will occur. In rapids the benthic invertebrates have a higher density than that in pools. In pools, the nekton and burrowing forms are denser in comparision to those in the rapids. The stream fish links the rapids and pools as it goes to pools for refuge and feeds in or at the base of rapids.

1. Organisms of the Rapids

The plants and animals inhabiting the rapids have unique morphological, physiological and behavioural adaptations to suit the currents which form the major limiting factor of the rapids.

(*i*) *Producers*

Phytoplankton is absent but it may be swept from the neighbouring bodies of standing waters. If it happens to pass through the rapids, it is soon destroyed as it is at the marcy of currents. In pool (slow moving parts of stream) however, it multiplies and becomes an important source of food. The producers of the rapids are the following:

(a) Filamentous green algae such as cladophora which grow attached to rocks.

(b) Encrusting diatoms which cover various surfaces.

(c) Aquatic mosses such as Fontinalis which cover stones.

(*ii*) *Consumers*

Zooplankton is absent. Some of the consumer animals are plant inhabiting forms which live attached to roots, leaves or stems of mosses and other producers. Some are rock inhabiting forms which inhabit the underside or exposed parts of submerged rocks. Certain animals have taken to burrowing in order to withstand current while the fishes are capable of moving against the water current. In general animals of the rapids show the following adaptations.

Streamline body. All animals from insect larvae to fish have a streamlined body which offers minimum resistance to water currents.

Flat body. Many animals of rapids have extremely flattened bodies to enable them to find refuge by clinging to under surfaces of stones and in crevices. Examples are stonefly nymphs and mayfly nymphs (1 and 2 respectively in Fig. 3.5).

Small size. Many fishes have a small size which enables them to move fast against water currents.

Sticky undersurface. Many animals like snails and flatworms have sticky undersurfaces by which they adhere to surfaces.

Hooks and suckers. These are fixation devices by means of which the animals of rapids grip the smooth surfaces like those of leaves and stems: For example black fly larva: simulium has a sucker at the posterior end and a headnet at the anterior end which is used for straining food from water (3 in Fig. 3.6). The blepharocerid larva has a row of ventral suckers (4 in Fig. 3.5) while the Caddis larva: Hydropsyche cements a net around itself for shelter and to be used as a food trap for animals and plants suspended in water (5 in Fig. 3.5). Some fishes like loaches have a sucker which represents a pair of modified fins.

Reduced gills of many fishes make them unfit for life in ponds where oxygen content is low. This adaptation is in response to high oxygen content of water in the rapids.

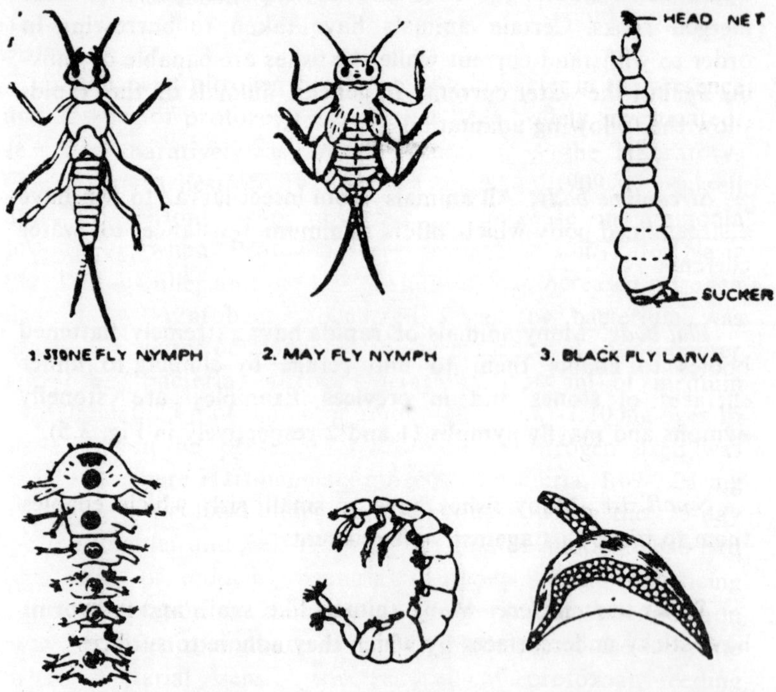

FIG. 3.5: Some representative consumers of the rapids.

Positive rheotaxis. (rheo=current, taxis=arrangement). This is an inherent behaviour pattern found in the animals inhabiting the rapids. These animals almost always orient themselves upstreams and if they are capable of swimming movements, they continuously move against the current. It is to be noted that animals of standing waters merely drift with the current and do not move against it.

Positive thigmotaxis (Thigma=touch or contact). This is also an inherent behaviour pattern found in the animals inhabiting the rapids. These animals have a tendency to clip close to a surface or to keep the body closely in touch with the surface. Thus if some stone fly nymphs are kept in a petridish they either lie in contact with the dish or cling to each other.

2. Organisms of the Pool

In the pools, the temperature of water is higher than that of rapids, the current is slow and the nature of bottom is different due to the accumulation of silt and decaying organic matter at the bottom. The pools harbour a variety of animals due to the rich nourishment they have.

(*i*) *Producers*

The producers are the phytoplankton transported from the neighbouring bodies of standing waters and backwaters, hence they show a mixture of planktonic forms.

(*ii*) *Consumers*

Zooplankton abounds in protozoans. The invertebrate animals consist of insects, fresh water muscles, snails etc. The vertebrate animals consist of fishes such as carps and car fishes reptiles, like crocodiles and turtles and amphihians.

SPRINGS

Springs differ from other habitats in the relatively constant chemical composition, velocity and temperature of water. They are said to be ecologist's natural constant temperature laboratory. This is because as the spring water is altered by photosynthesis and respiration, it passes downstream and is replaced by water having the original composition from underground. In the spring therefore, the communities can be studied under known conditions. Knowledge is available about the following springs:

1. *Hot springs* with high salinity found in volcanic areas such as in Iceland, New Zealand, U.S.A. and Africa.
2. *Hard water springs* (having average temperature) found in the limestone districts of Florida, Denmark and Germany.
3. *Soft water springs* which emerge through sandstones, crystalline rocks and shales.

In the hot springs with high salinity, as one goes downstreams the temperature and salinity decrease steadily. The community composition changes accordingly, hence study of communities of a spring run shows a natural ecological gradient. The study of hot springs have established the upper limit of temperature tolerance for the following organisms.

Blue green algae	88°C
Protozoa	54°C
Insects	50°C
Fishes	50°C
Bacteria	88°C

The study of springs shows that many species occur in a particular region just because the springs provide them with a favourable environment. For example arctic insects occur in German springs due to their low summer temperature, similarly the hot springs of iceland harbour many organisms which normally occur only in the regions of warm climates. It seems that springs have offered refuge to aquatic organisms during geological periods when climate changes were taking place.

Chemical Condition of the Bottom Soil in the Aquatic Media

A well aerated pond is a relatively self-supporting ecosystem for production of nutrients. Total and available quantity of raw materials, suitability of environmental conditions and supply of nutrients are other vital factors influencing productivity of that ecosystem. Some major chemical factors influencing productivity are:

Hydrogen Ion Concentration. pH of the substrate soil governs the adsorption and further release of ions. The adsorption and release both take place at the soil-water interface. The soil becomes acidic in a stagnant oxygen deficient pond. This increased acidity inhibits the bacterial action thus suppressing the continuous decomposition of organic material. Seldom, the soil is naturally buffered.

Chemical Systems. It appears there are certain established chemical links and systems which are delicately balanced to promote natural productivity. Any tilt in these complicated long chains will affect the productivity schedule. Some of the well identified systems are dealt with to indicate involved factors. Bottom soil substrate influences the chemistry of water in many ways. Some delicately balanced chemical systems exist in the soil. The type of system depends upon availability of minerals and pH of waters. Only the common well-identified systems are treated here.

Iron-sulphur-phosphorus System is one in which H_2S is formed at the bottom by the anaerobic reduction of sulphates by bacteria. H_2S in turn reduces ferric ions to ferrous ions. The products liberated during these decomposition are rapidly converted by bacteria and iron (Fe) is deposited as $Fe(OH)_3$ (Ferric hydroxide).

Calcium Carbonate-phosphorus System. This system exists in ponds having alkaline water and higher calcium content. In such cases most of the added phosphates are precipitated with calcium. In these ponds phosphorus is quickly adsorbed by the bottom mud.

Iron Humus-phosphorus System. Bottom humus contains humic acids. But the exact composition of humic acids is not so well understood. Humic acids form complexes with heavy metals by chelation. In an aerated pond, iron is in ferric form and forms large particles. The iron linkage of humic acids is counteracted by calcium concentration. It is believed that such large quantities of iron are responsible for highly coloured (blue) waters. Chelation plays a significant role in lending about the deeper blue-green colour.

Iron or Aluminium-silicon-humus System. The linkage of phosphorus and silicon to iron and aluminium—humic acid complexes is known to occur in some ponds.

Phosphorus. Phosphorus is necessarily taken up by bacteria and algae. The uptake and liberation follow a definite well organised cycle. In the presence of bacteria amount of phosphorus remaining in water at equilibrium is more than if bactericides like antibiotics are used.

The total amount of dissolved phosphorus present in the pond is not so important to aquatic productivity as the amount of phosphorus freely available. Phosphate ions form insoluble compounds with iron and aluminium in an acidic environment and form compounds with Ca in an alkaline environment. Phosphorus occurs in both organic and inorganic state. Organic compounds containing phosphorus are, phytin and its derivatives, phospholipids and nucleic acids. Remaining phosphates are adsorbed by the mud, resulting in precipitation.

Phosphorus is liberated to epilimnion. During decomposition of littoral vegetation this free phosphorus is taken up by phytoplankton and littoral vegetation. Phytoplankton steadily looses phosphorus in the form of water soluble compounds. These compounds ionise and exist as free ionic phosphates. Bacterial putrefaction of the phytoplankton results in senimentation. Phosphorus is liberated from the sediments to hypolimnion and settles in the sludge or soil-water interface. Remaining phosphorus from the sediments slowly diffuses into water.

Nitrogen. Organic matter settled on pond bottom is broke n down by a number of micro-organisms. End products of this decomposition are carbon dioxide, water and ammonia. Ammonium salts are formed from ammonia, which by bacterial action are converted into nitrates. Denitrifying bacteria reduce the ammonium salts and nitrates to free nitrogen. Bottom soil adsorbs remaining NH_4 ions. Extent of adsorption depends on the concentration of ammonium salts already in the soil. Some ammonium salts and nitrates being highly soluble are leached out with the percolating water. Ammonia released during decomposition evaporates in high temperature. Main source of nitrogen to the pond is bottom soil. But its efficiency is often in doubt as some recommend to add artificial fertilisers to the pond.

Organic Carbon and C/N Ratio. Percentage of organic carbon is an important factor in fisheries development. Though it cannot be stated definitely, in practice it was observed that 0.5-1.5 per cent organic carbon yields average fish production while 1.5-2.5 per cent is optimal. A higher percentage is not desired for better productivity. Same holds true for C/N ratio. A ratio less than five indicates poor production, more than 15 also is less favourable while 10-15 ratio is optimum for good production.

Potassium. Potassium is always present in water and being necessary for submerged weeds is quickly taken up by them. But the optimal concentration is not so well ascertained.

Calcium. Calcium is important for its role in translocation of carbohydrates. It is also a vital integral component of plant tissue system. Calcium reduces toxic effects of single salt solutions of other elements and increases availability of other ions.

There are certain micronutrients which even in minute quantities promote metabolism of plants and animals. Such naturally occurring substances are called "trace elements" which impart significant differences in the productive capacity of fresh waters. Indeed, they play a vital role in the bioenics *i.e.* the economics of any biotic community. It appears that manganese (Mn) promotes photosynthesis, heterotrophic growth of phytoplanktonolgal population in particular and increase in crustacean fauna.

4

Trophic Relationship in Protozoa

Free-living Protozoa occupy a range of trophic levels. Those which feed on algae, both the unicellular and filamentous varieties, are primary consumers in the herbivore food chain dependent on fresh autotrophic organic matter. Other protozoans exploit bacteria as a food source and as microobivores perform a role in the food chain based on dead organic matter. Both these trophic groups of Protozoa, and other small elements in the micro-and meiofauna, are exploited by predacious protozoan species. The Protozoa form an element in the larger micro -and meiofaunal communities of soils and aquatic environments, but on a smaller scale the various protozoan groups themselves constitute a community with a defined trophic hierarchy (Figure 4.1).

The decomposition food web based on dead organic matter is extremely important in nature, because it is responsible for recycling carbon, phosphorus, nitrogen, sulphur and other essential elements in the living world. The pool of dead organic matter which the decomposers exploit, is derived from primary production and is composed of dead plant tissue, the faeces of heterotrophs, their exudates and exuviae and body remains where death is non-predatory. Thus two basic food chains can be distinguished in nature, one based on fresh plant tissue and the other on dead organic matter, the former providing an energy source for the latter; the decomposers recycling the essential nutrients required for photosynthesis.

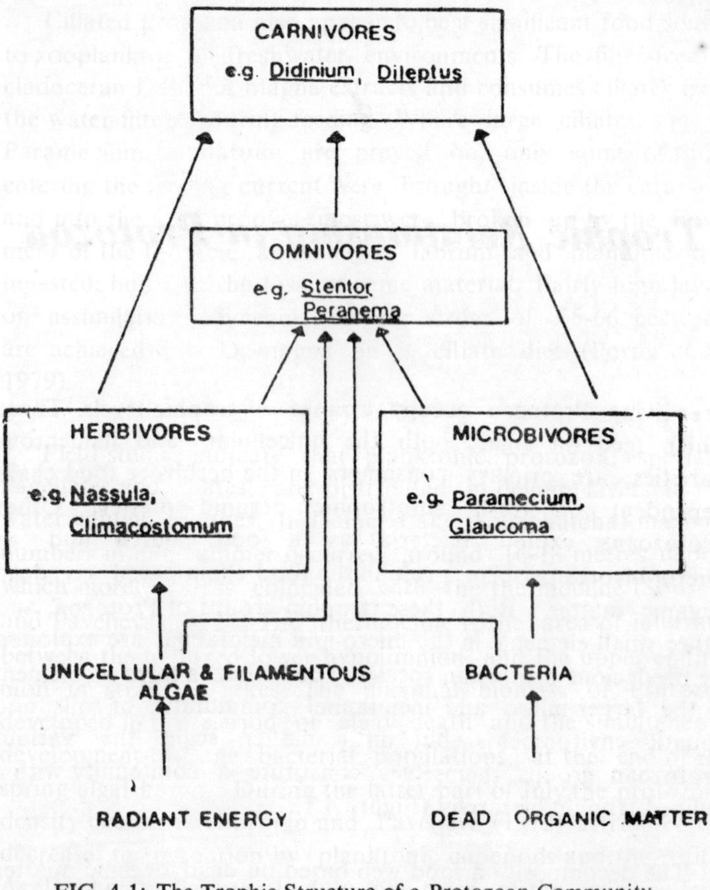

FIG. 4.1: The Trophic Structure of a Protozoan Community.

The majority of Protozoa are microbivores exploiting the decomposer bacteria as a source of energy, transforming some of the assimilated bacterial production into protozoan tissue and some into energy for metabolic processes and movement. The Protozoa in turn become potential energy for a variety of predators including carnivorous Protozoa, copepods and other meiofauna. The environments inhabited by protozoans are often heavily dependent on the decomposer food chain. Many freshwater and estuarine aquatic habitats, particularly the benthic zones, are part of so-called subsidised ecosystems. These are ecosystems, which although performing endogenous primary production, much of which eventually finds its way into the pool of dead organic matter (*i.e.*, autochthonous dead organic matter), also receive an input or subsidy of dead plant tissue from surrounding terrestrial or semi-terrestrial ecosystems (*i.e.*, allochthonous dead organic matter). Soils and litter layers are also important protozoan habitats and represent the area of terrestrial ecosystems where the decomposition processes are performed.

The obvious ecological role of microbivore Protozoa is the transfer of bacterial production through protozoan cellular material as potential energy to successive trophic levels. However, it is now becoming increasingly clear from studies on nutrient recycling and bacterial growth dynamics, particulary in the soil, that the contribution made by Protozoa may be greatest and more complex than we had previously assumed. There is now evidence that Protozoa may stimulate bacterial growth and hence the rate of decomposition by various means. In addition it seems that Protozoa play a significant role in the recycling of essential nutrients, especially carbon and phosphorous.

The efficiency and rate at which protozoans transfer energy from one trophic level to another, is determined from physiological studies on species populations. The physiological data on growth, respiration and feeding are transformed into units of energy, so that the energy intake and partitioning of energy within the organism can be considered. This energetics approach integrates the various aspects of an organism's physiology, so that the overall functioning of an organism as an element in a population or a community can be ascertained. The flow of energy through a species or population can be expressed simply in the energy budget equation, which was proposed by Ivlev (1938):

$$C=P+R+F+U$$

where C=the energy ingested as food; P=energy produced as new tissue and reproductive products; R=heat dissipated during respiration or respiratory energy loss; F=egested energy; U=nitrogenous excretion.

$$A=P+R$$

where A=energy assimilated or utilised.

In protozoans P is solely growth in the normal asexual life-cycle, because the whole body is the reproductive product, passed on usually as two new individuals to the succeeding generation. Egestion and excretion are for obvious reasons logistically difficult to quantify in such small organisms, so these are usualty calculated by subtraction from the other empirically-determined parameters of the energy budget.

The efficiency by which an organism uses the energy it ingests for production and metabolic function is expressed by a series of ratios which are termed coefficients of efficiency:

$$\text{assimilation efficiency}=\frac{A}{C} \text{ or } \frac{P+R}{C}\times 100$$

The indicates what proportion of the ingested food or energy is actually used by the organism for production and respiration.

$$\text{gross production efficiency}=\frac{P}{C}\times 100$$

which gives us a crude index of how much of the consumed energy appears as production.

$$\text{net production efficiency}=\frac{P}{A}\times 100$$

This indicates the proportion of assimilated energy which is directed into growth and reproduction. If the proportion of the assimilated energy going into production is known, then it follows that the proportion which is dissipated as heat during respiration can be derived by subtraction.

These various coefficients of efficiency are very valuable when making comparisons of organisms from the same trophic level, and for comparisons between trophic levels. As a community level the components of the energy budget expressed in joules or calories, and the coefficients of efficiency, give us an insight into the overall energy flow and energetic interactions between organisms.

Protozoan Energetics

Studies on protozoan energetics are relatively recent, although the physiological ecological approach to the functioning of organisms developed in the 1940s, and gained momentum, so that by the 1960s and 1970s a wealth of information was available on the energetics of invertebrates and vertebrates in the ecological literature. The first energy budget produced for a protozoan was for the amoebae Acanthamoeba spp. (Heal, 1967). This energy budget, valuable through it was in leading the way for subsequent protozoan energetics studies, had its limitations, since it applied to one temperature only and suffered from a lack of calorimetric data. Later investigations attempted to integrate the variable of temperature, so that energy budgets could be applied to field populations. The aim of energetics studies is to elucidate the physiological performance of an organism in relation to its food source and its predators in the natural environment. In this respect, protozoan ecology lags behind that of many other invertebrate groups, particularly macrofaunal species. The area is beset by problems, not least of which is our knowledge of the nutritional requirements of most protozoan species. They can be grown in the laboratory in axenic and monoxenic culture, but little is known about protozoan feeding preferences in the wild, or about how the elements of competition for food on an intra- and interspecific basis affect energy consumption and other aspects of the physiology. As Lee, J.L. (1980) points out, our understanding of energy transformations in food webs involving Protozoa require an elucidation of two components related to food quality: first, information on the molecular constitution of the prey, and secondly, the ability of Protozoa to recognise and utilise it. Thus while we are able to demonstrate how food quality and quantity affect growth and reproduction in many species, we can only place this information into the context of the natural community to a limited extent.

The reasons for the paucity of information on protozoan energetics are several. Until recently the smaller animals in the ecosystem were mentioned only in passing, if at all, and dismissed as probably contributing little to the overall functioning and energy flow of aquatic and terrestrial communities, and hence there was no need to study them in detail. Protozoan physiology and calorimetry are faught with technical difficulties imposed by the small size of these organisms. The ideal invertebrate for an energetics investigation is a macrofaunal monophagous herbivore, with a one-year life-cycle and recruitment at one specific period only. The Protozoa are neither macroscopic, nor monophagous on one species of bacterium or alga, and they have continuous recruitment over a short period. Furthermore, they frequently live in communities with high species diversity.

(i) The Energy Equivalents of Protozoa and their Food

In order to compose a picture of energy flow through the physiological function of a population or community, it is necessary to acquire data on the energy value of organisms and their food. The energy content of such materials is usually determined by means of microbomb calorimetry (Phillipson, 1964).

Table 4.1 shows the energy values of a range of protozoan species derived from a number of studies. The energy values derived for various ciliate species do not differ radically, and it is not unreasonable to apply such values as conversion factors to other ciliate species. The low value for Noctiluca is attributable to a relatively high volume of cell sap and probably a water-bound component included in some of the samples (Finlay and Uhlig, 1981). There is obviously a need for more detailed information on flagellates and on amoebae where data are limited. The energy content of protozoan cells is within the range reported for a wide spectrum of invertebrate organisms showing energy values within the range 17.58-28.46 J mg^{-1} ash-free dry weight (Prus, 1970).

TABLE 4.1

The Joule-equivalent Values of Protozoa Derived from Bomb Calorimetry

Species	*J mg⁻¹ Ash-free Dry Wt*	*Author*
Ciliate		
Tetrahymena pyriformis	24.85±0.84	Slobodkin & Richmann (1961)
Tetrahymena pyriformis	21.3 ±0.3	Finlay & Uhlig (1981)
Tetrahymena pyriformis	19.8 ±0.2	Rogerson (1979)
Paramecium caudatum	23.5 ±0.7	Finlay & Uhlig (1981)
Euplotes spp.	19.3 ±0.7	,,
Colpidium campylum	20.15±1.5	Laybourn & Stewart (1975)
Sarcodina		
Amoeba proteus	17.51±0.3	Rogerson (1979)
Mastigophora		
Ceratium hirudinella	23.2 ±0.8	Finlay & Uhlig (1981)
Noctiluca miliaris	6.6 ±0.6	,,

The bacterial food sources of many protozoans show variable energy values. Laybourn and Stewart (1975) report Moraxella isolated from fresh water as having an energy content of 20.36 ± 0.4 J mg^{-1}. Other freshwater species isolated from Paramecium were found to have an energy content of 21.0 ± 0.4 J mg^{-1}, while marine bacteria have a calorific value of 5.5 ± 0.8 J mg^{-1} (Finlay and Uhlig, 1981). The lack of information on the food of protozoans emphasises the problems outlined earlier.

(*ii*) *Energetics*

Complete energy budgets related to variables such as food concentration and temperature are few, although there are more examples of what may be described as partial energy budgets covering feeding and production. Those complete budgets which are available do provide some insight into the overall efficiency of exploitation and use of resources by protozoans. Table 4.2 shows some energy budgets for two very different protozoans feeding on Tetrahymena in an experimental situation. One of the species, Amoeba proteus, moves along a substrate engulfing the food it encounters, while the other species, Stentor coeruleus, is a large, mainly sedentary ciliate which actively draws water over the oral region where it filters out food items. In each case a single Amoeba or Stentor was placed in a volume of 500 μl of medium with differing concentrations of prey. In the case of Amoeba the density of the prey was considerably greater than that available to Stentor. The higher prey concentrations offered to Amoeba may not reflect the natural situation, but do result in higher levels of consumption. Indeed the level of consumption continued increasing up to prey densities of 4,000 per Amoeba (Rogerson, 1981). Given the much lower densities of food available to Stentor, comparatively high levels of consumption were achieved, epecially when one considers that the energy budget for Stentor spans a shorter period because the generation times are shorter than those in Amoeba (see Table 4.2). In other words the rate of consumption is apparently higher in Stentor on lower food densities than in Amoeba proteus. This disparity is a function of the mode of feeding in each species. Stentor filters the medium in which it lives for food. The rates of filtering which can be achieved by ciliates are high. It has been estimated that a ciliate can clear a volume of water equal to 20,000 times its

TABLE 4.2

Generation Energy Budgets for Protozoans. Food Source Tetrahymena Pyriformis. Parameters of the Energy Budget Expressed as $\mu J \times 10^3$. See Text for Explanation of Abbreviations

Species	*Gen. Time hrs*	*Temp °C*	*Food Ratio*	*C*	*P*	*R*	*F+U*	*A*	*P/A%*	*P/C%*	*A/C%*
Amoeba proteus	111	15	125:1	4.68	1.78	0.59	2.30	2.37	75	38	50
,,	68	15	2,000:1	11.21	2.53	0.61	8.06	3.14	80	22	28
,,	83	20	125:1	4.46	1.23	0.66	2,56	1.89	65	27	42
,,	48	20	2,000:1	5.88	1.75	0.45	3.63	2.25	78	30	38
Stentor coeruleus	43	15	30:1	2.93	2.21	0.06	0.66	2.26	97	75	77
,,	40	15	50:1	4.26	2.71	0.06	1.48	2.77	97	63	65
,,	42	20	30:1	3.04	2.17	0.09	0.77	2.66	96	71	74
,,	39	20	50:1	4.39	3.23	0.12	1.03	3.35	95	74	76

own volume in one hour. Put in perspective, this is 20 times higher than lamellibranch molluscs (Fenchel and Jorgensen, 1977). Amoeba proteus has a less efficient means of acquiring energy and depends upon random contact with its prey as it moves along, and thus much lower rates of consumption are achieved in relatively high prey densities.

Having consumed energy, an organism then extracts and utilises the energy available to its digestive capabilities from the food, and partitions the assimilated energy into production and maintenance. The efficiency by which energy is derived from ingested food is given by the assimilation efficiency (A/C). The figures in Table 4.2 show that Stentor possesses higher assimilation efficiency, taking more of the available energy from the same type of food than Amoeba. The reasons are difficult to pin-point, but the differences may be a function of a more efficient system of digestion and absorption in Stentor. This species has a higher overall production and higher rates of production than Amoeba, which is apparent from the quantity of μJ produced and the difference in the generation times. Put another way, Stentor grows to a larger size in a shorter time and produces more generations in a given time than Amoeba. The efficiency of net production (P/A) is higher in Stentor, but both species are efficient producers. In each case a large proportion of the assimilated energy is directed into production, and respiratory costs are low. The very low levels of respiratory energy loss in Stentor are probably attributable to its largely sedentary habit. These carnivorous protozoans possess net production efficiencies which are considerably higher than those found in metazoan carnivores. A review of production efficiencies in ectothermic metazoans by Humphreys (1979) indicates that non-social carnivorous insects achieve net production efficiencies around 56 per cent, while other metazoan invertebrate carnivores are lower, averaging 28 per cent.

Net production efficiencies among the small number of bacterivorous Protozoa which have been investigated show considerable variation. Strombidium sp., Tiarina fusus and

Diophrys sp. achieve levels of efficiency of 2-6 per cent, 15-42 per cent and 14-40 per cent respectively at temperature ranging between 22°C and 25°C (Klekowski and Tumantseva, 1981), while another ciliate, Spirostomum ambiguum, has a net production efficiency of 55 per cent (Klekowski and Fischer, 1975). The bacterial feeding flagellates Ochromonas sp. and Pleuromonas jaculans are capable of net production efficiencies around 60 per cent at 20°C (Fenchel, 1982). Strombidium would appear to be exceedingly inefficient, but this may be a function of the experimental temperature; it may be a species better adapted to lower temperatures. The other species are capable of efficiencies within and above those reported as characteristic of metazoans (Humphreys, 1979). However, it should be noted that Humphreys (1979) did not include bacterivores among the trophic groups in his review, probably because the literature on such feeders is sparse. The efficiency at which bacterivorous protozoans convert assimilated energy into production is comparable to that displayed by the bacterial feeding nematode Plectus palustris, which is capable of net production efficiencies in the range 51.2—73.7 per cent at 20°C, varying in relation to developmental stage and food concentration (Schiemer *et. al.*, 1980).

It is interesting to speculate on how the relationship between respiration, production and assimilation, and consequently the efficiency of energy utilisation by an organism, has changed during the course of evolution. Phillipson (1981) suggests that within multicellular ectotherms production efficiency fell progressively during evolutionary advancement, while maintenance costs, as a proportion of total assimilation, increased. The development of homeothermy was accompanied by a further fall in production efficiency and an associated increased in maintenance costs. At present there are insufficient data on unicellular organisms to draw any firm conclusions, but the high net production efficiencies displayed by many of the protozoans so far studied would suggest that Protozoa may be among the most efficient heterotrophic organisms, and that the evolution of a multicellular level of organisation and associated morphological and physiological complexity, has resulted in a decrease in the efficiency of production.

Production studies related to food consumption give us a gross estimate of conversion efficiency (P/C) of ingested energy into production or yield. The gross production efficiency is a useful ecological index of the quantity of energy transferred as yield by an organism from the trophic level it exploits as food to its own predators in the trophic level above it. Those bacterial and fungal feeding Protozoa which have been studied in this respect have varying gross production efficiencies. Heal (1967) found Acanthamoeba sp. fed on yeast cells to attain a gross conversion efficiency of 58 per cent at 25°C. Tetrahymena pyriformis grown on bacteria achieved a gross production efficiency of 50 per cent at the same temperature (Curds and Cockburn, 1968). Higher levels were achieved by Colpoda steini where gross production was 78 per cent of consumption (Proper and Garver, 1966), while lower values of P/C ranging from 3 to 11 per cent over 10-20°C have been reported in Colpidium campylum feeding on bacteria (Lay-bourn and Stewart, 1975). The bacterivorous flagellates Ochromonas sp. and Pleuromonas jaculans achieved yields of 34 per cent and 43 per cent at 20°C respectively (Fenchel, 1982). The wide variations on these gross production efficiencies for Protozoa are probably partly a function of the temperature at which experiments were conducted—generally the efficiency of production declines with decreasing temperature—but other factors may be involved. Production may also vary in relation to food concentration, because the quantity of available energy may limit the rate of ingestion, but in addition the energy value of the various bacterial and fungal food sources may vary. We have only limited information available to us on the energy content of these materials.

Of necessity, the foregoing energetics studies have been performed in the laboratory, but production studies for field population have been attempted, usually combining aspects of laboratory-determined energetics with field data in a number of ways. One of the earliest studies which adopted this approach was carried out by Fenchel (1975) on the arctic tundra pond. Fenchel estimated carbon flow through the most important groups of Protozoa in the pond, giving a minimum estimate of standing stocks as well as transfer rates. The rates of protozoan digestion of bacteria were determined in the laboratory at a series of temperatures. The digestion rates were multiplied by the number of food vacuoles in freshly collected Protozoa to give an estimate of grazing rates in the field. Zooflagellates were quantitatively the most important protozoan group. During 24 hours in the summer months the zooflagellates consumed 15 mg Cm^{-2}. The calculations showed that the zooflagellates consumed a number of bacteria equal to their body weight in 24 hours, whereas ciliates grazing on bacteria ingested the equivalent of 25 per cent their own biomass in the same period. These rates of bacterial grazing are high and serve a valuable role in the saprovore food web.

In the benthos of many temperate lakes ciliates tend to be the dominant protozoan group. Finlay (1978) estimated production and respiration for the ciliate community of a small eutrophic unstratified lake in Scotland, by combining data on the reproductive and respiration rates of a range of ciliate species with seasonal field data on numerical density seasonal and species composition. The patterns of production and respiration decreased with increasing depth, Figures 4.2 and 4.3. Mean production ranged from 40 to 345 J cm^{-2} year ^{1}, while respiration ranged from 4 to 16 cm^{-2} $year^{-1}$. These high levels of production are at their maximum in the summer months. In stratified eutrophic lakes the picture is quite different, for here production is depressed in the summer by the conditions of low temperature and anoxia which prevail in the benthic zone (Figure 4.4). The differences in production between benthic protozoan communities in stratified and unstratified lakes is obvious from a comparison of Figures 4.2 and 4.4. Studies have shown that many of the ciliates in the benthic zone of stratified eutrophic lakes migrate

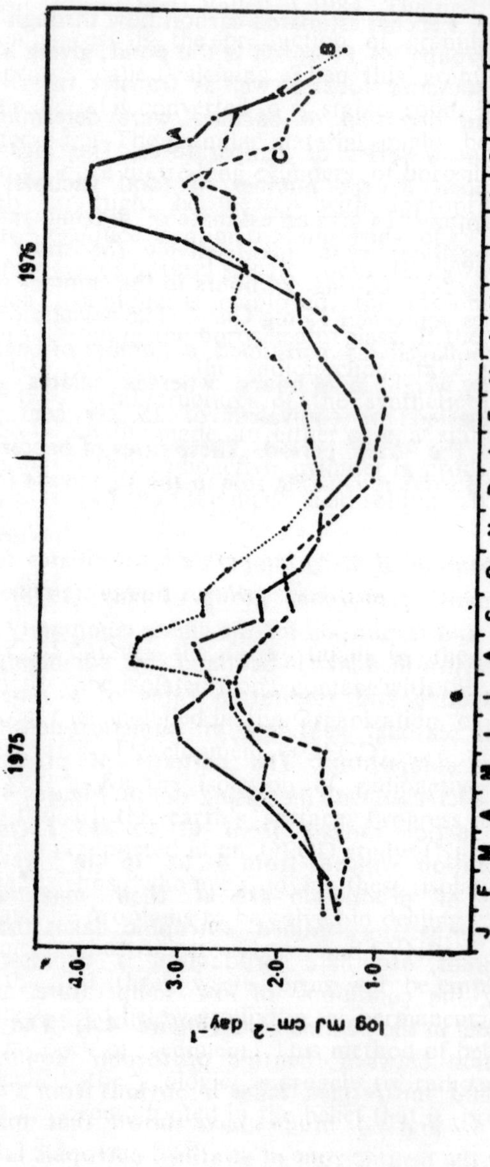

FIG. 4.2: Production by the Ciliate Community of a Small Eutrophic Unstratified Lake at Three Sites (A, B and C) of Increasing Depth, A Being the Shallowest Site.

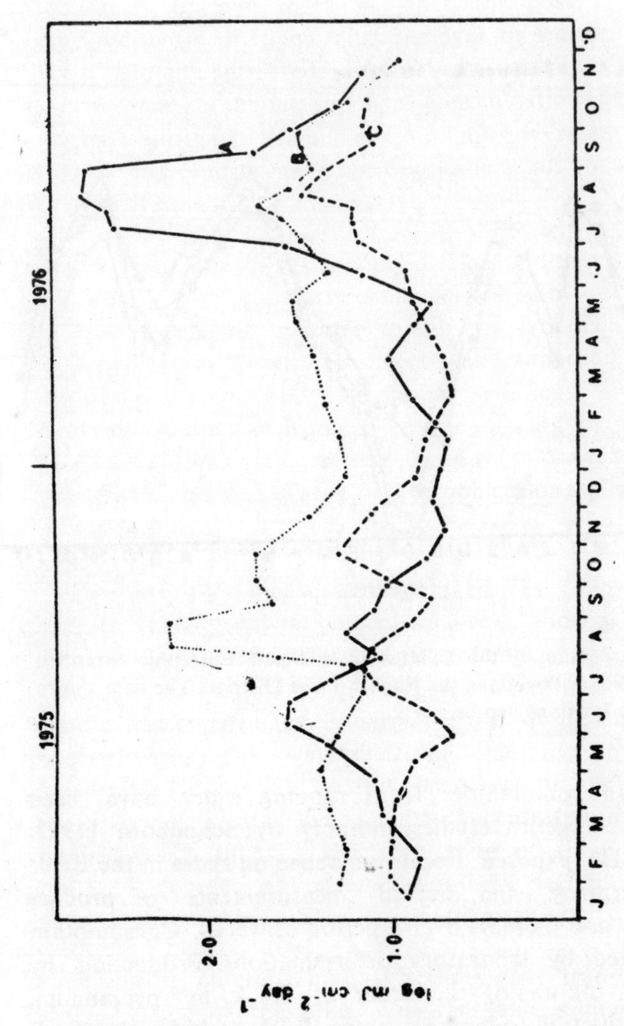

FIG. 4.3: Respiration by the Ciliate Community of a Small Eutrophic Unstratified Lake at Three Sites (A, B and C), A Being the Shallowest Site.

vertically in the water column out of the anoxic conditions of the sediment surface and lower waters (Bark, 1981; Finlay, 1981).

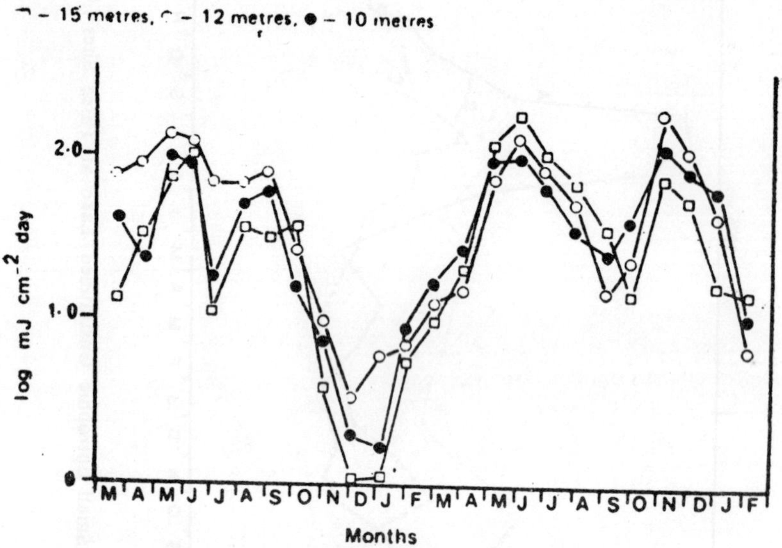

FIG. 4.4: Production by the Ciliates of a Small Eutrophic Stratified Lake. Esthwaits in the English Lake District. The data shown are for three depths.

Protozoan communities from running water have been subject to energetics studies, notably by Schonborn (1977, 1981a, b). He exposed testate amoebae on slides in the River Saale in Germany, and derived measurements of production, density and biomass over a period of weeks. Consumption was estimated by laboratory determinations. Production by Testacea in the was of the order of 2.6 kJ m^{-2} per annum, giving a production-to-biomass ratio (P/B) of 16.8. The levels of consumption were estimated as 2,444 mg diatoms m^{-2} $year^{-1}$ (Schonborn, 1981a). In another study on ciliated Protozoa in a small brook, annual production was 4,219 mg m^{-2} with an extremely high production-to-biomass ratio of 195

(Schonborn, 1981b). The ciliates achieve high levels of consumption, ingesting 14.8g m^{-2} of bacteria, diatoms and other Protozoa, the largest portion being bacteria.

A river environment is not an optimal one for many Protozoa and one may reasonably expect standing water communities, particularly those tending towards eutrophy, to be more productive. This is borne out by a comparison of the available figures for annual production in lakes and rivers. The production achieved by ciliates in Schonborn's (1981b) study of a river are two or three orders of magnitude lower than those reported for lake ciliates in an unstratified eutrophic lake by Finlay (1978) and in Esthwaite, a dimictic eutrophic lake. The physical and chemical conditions of standing and running waters differ, and this is reflected in the energetics of communities in each type of aquatic ecosystem.

(*iii*) *How Protozoa Maximise Net Energy Returns from their Feeding Behaviour*

The acquisition of food or energy requires some expenditure of energy in terms of searching and handling food. For the most efficient functioning an organism must maximise the net energy return from its feeding activity. Optimal foraging theory has been applied to Metazoa, particularly carnivores, but the concept can be applied in a modified from to other types of feeders.

Protozoa fall into different trophic cotegories. Many ciliates and flagellates are essentially filter feeders, drawing the liquid medium over a specific area of the cell and extracting food from the water current. Many other Protozoa, particularly amoebae, are deposit feeders, engulfing particles of food with which they make contact. Carnivorous Protozoa vary in their feeding behaviour; some, such as Didinium, actively hunt, but capture is random. The ciliate does not actually pursue a prey, it depends on random contact. Other carnivores, for example the suctorians and representatives of other groups, are ambush predators, waiting for their prey to come to them. Much of the current optimal foraging theory may not be applicable to Protozoa, but since there is evidence of food selection by Protozoa, there is scope for considering how some of these single-called organisms maximise energy return from their feeding behaviour.

Lam and Frost (1976) and Lehman (1976) have applied the principles of optimal foraging to filter feeders. In carnivores it is necessary to consider energy yield per unit handling time of different prey items with differing energy content in different densities, in relation to searching time. In filter-feeding organisms it is the particle density in the medium, the size of the particles in relation to particle selection, and the rate of food processing which are important. Here the filtering rate of the medium by the organism corresponds to the searching time in carnivores. In filter-feeding metazoans, such as bivalves or filter-feeding copeponds, the net rate of energy gain equals the rate of energy assimilation, minus the cost of filtering, minus the cost of rejecting unsuitable particles. Townsend and Hughes (1981) have summarised the principles of maximising energy return in filter-feeding invertebrates:

$$Q = E_a - E_f - E_r$$

where Q (=E/T)=net rate of energy gain; E_a=rate of energy assimilation; F_f=cost of filtering; E_r=cost of recycling unwanted particles.

Each component of the equation is complex. The rate of energy assimilation (E) is a function of the proportion of particles of any given type which are acceptable for ingestion, the density of any given particle type in the water, the energy value of the particle and the filtering rate. The cost of filtering (E_f) has to take into account the energy expenditure as a result of drag on the filtering apparatus. Rejection energy costs (E_i) are a function of the proportion of any particle type rejected and their density in the medium in relation to the filtering rate.

Lehman (1976) plotted filtering rates as a function of particle density. A maximum filtering rate is achieved at a particular particle density of a given particle type; the maximum corresponds to the density of the particle when the gut first becomes full, and thereafter lower filtering rates maintain the gut in a full condition. In nature animals are not usually confronted by one type of particle only, but an array of particles, some of which are suitable for ingestion and some which are not. Those which fall into the category of being suitable as food may have a different energy content. The energy value of any given type of particle may be a function of its size or its digestibility. In order to maximise the energy return from feeding, those particles with the highest energy yield should always be accepted, when their density in the array of particles filtered is proportionally high. However, as the density of energy-rich particles declines in the overall volume of particles being filtered, rejection costs become high relative to the energy gained. When rejection costs are high in filter-feeding metazoans, even non-digestive items should be ingested and passed through the gut along with energy-yielding material (Townsend and Hughes, 1981). The organism must of course be capable of recognising and grading the particles it encounters in terms of their energy value.

Theoretically the same principles are applicable to filter-feeding protozoans. Although there is information available on how feeding currents are produced in ciliates and how much food is consumed per unit time, our knowledge of filtering rates in relation to particle size and concentration is limited. In a number of respects Protozoa appear to be much simpler than many metaozoan filter feeders. Many bacterivorous ciliates appear to discriminate what they ingest on the basis of particle size and shape only, and this is related to their mouth morphology (Fenchel, 1980d). In other words they apparently ingest all particles in the size range within the retention capacity of the oral cilliary organelles, irrespective of the energy value of the particles. There is evidence, however, that some ciliates, such as the large omnivorous species Stentor, do discriminate the type of particle ingested. In ciliate bacterivores Fenchel

(1980b, d) showed that each species has its own size range of particle, there being a correlation between the minimum particle size retained and the free space between adjoining cilia of the ciliary filters. Rejection occurs in the sense that particles outside the size repertoire of a given species are not retained, so that some energy cost is implied.

It appears that the filtering rate in ciliates is constant irrespective of particle density in the medium (Fenchel, 1980b). The uptake of particles is a linear function of particle density increasing, experimentally, as a linear function of time. The maximum uptake rate is determined by the rate at which particles can be phagocytosed. As particle density in the medium is increased, the filtering apparatus becomes clogged by particles, because the rate at which ingestion can be achieved is limited; thus some of the retained material will be lost from the ciliary apparatus (Fenchel 1980a, b). The energetic cost of overcoming the resistance or drag to the ciliary filter, which presumably is constant since the filtering rate is constant, has been estimated as being very small, of the order of less than 1 per cent of the energy budget of a small ciliate like Cyclidium (Fenchel, 1980b). Added to this of course is the energy cost of bringing about ciliary movement.

Do these comparatively simple organisms attempt to maximise energy return from their feeding activity as do many metazoans? The answer is apparently not in most of the ciliate species for which detailed information on feeding activity is available. Any organism which filters at a constant rate irrespective of the density of particles in the medium is energetically inefficient. To be efficient an organism must modulate the filtering rate so that in any given particle density the quantity of material passing over the filtering apparatus to the mouth is the maximum which can be adequately coped with. A protozoan with an apparently constant filtering rate will, in high particle densities, receive too much material to phagocytose and blocking of the filter occurs. Similarly it could be argued that any organism which indiscriminately ingests all particles in a particular size range, is not maximising energy return. A system of recognising and accepting only that material in the array available which gives a good net energy return, is cleary energetically efficient.

Protozoa are very efficient energetically, having high growth rates and low respiratory costs in general. Perhaps there is no real advantage to a unicellular, efficient animal living in a relatively rich food environment in evolving or pursuing a 'strategy' designed to maximise energy return. Such 'strategies' may have evolved in the ecological interactions of higher organisms, in response to longer life-cycles and greater physiological complexity imposing pressure to 'conserve' energy. The size of particles retained by ciliates appears to be fairly small in range. For example, bacterivore species most efficiently retain particles of 0.3-1 μm (Fenchel, 1980a), so that there is a high probability that most of the material ingested will be bacteria, although of course not all the bacteria will be digestible or provide a good net energy return.

In other respects ciliates do reduce the energy costs involved in feeding, in that several species have been shown to flocculate bacteria, thus concentrating the food source and rendering it more accessible to grazing.

Ciliates are among the most efficient Protozoa. Less efficient groups may need to modify their feeding behaviour to maximise energy returns. As yet our knowledge of the intricacies of protozoan feeding behaviour under conditions which resemble those of the natural habitat is very limited. We have a wealth of information on the ultrastructure of the structures involved in feeding, and much data on how energy is ingested in relation to environmental physical and chemical conditions, but this is insufficient to consider optimal foraging theory as it may apply to the various protozoan trophic groups. It is an area of protozoan ecological energetics which should be explored, because it sheds more light on an organism's functional interactions with its food sources and on intrinsic energetics

r-K Selection in Protozoan Populations

During evolution, adaptations conferring competitive ability are believed to have evolved with a resulting loss in the innate or intrinsic rate of increase (r_m). This hypothesis, referred to as the r-K selection theory, has been widely applied in animal ecology since it was first proposed by MacArthur and Wilson (1967). Any population in any given environment will have its own characteristic survival rate, growth rate and reproductive rate. Each of these parameters will be partly innate and partly a result of prevalent environmental factors. The ability of an organism to increase, that is its innate or intrinsic capacity for increase (r_m), can be defined under a given set of conditions. Ecologically the hypothesis relates the inner characteristics of the organism to the physical, chemical and biological features of the environment. It has been suggested that some groups of organisms can be recognised as being r- or K-selected (Pianka, 1970).

r-selected organisms are subject to density-independent population control, which results in the population continuously increasing, but rarely reaching its limit. Among Metazoa, such organisms usually have high values for r_m, rapid development, reduced body size, normally a single phase of reproduction, a short life-cycle and reduced competitive ability. Organisms which are K-selected are subject to density, dependent regulation, resulting in the population usually reaching its saturation density. Typically, K-selection involves slow development and a delay in reproduction, which is repeated, together with a large body size and a long life-cycle, usually exceeding one year. Intra- and interspecific competition is maked in K-selected animals. The hypothesis rests on the principle that selection acts reciprocally on the intrinsic rate of increase (r_m) and the population saturation density (K). Thus r-selection under continuous growth should increase r_m, but reduce saturation density (K), whereas K-selection reduces the value for r_m but increases competitive ability and K.

The r-K selection hypothesis has recently been tested on protozoan populations. Luckinbill (1979) set out to examine the effect on the saturation density (K), of selection for higher growth rate, maintaining that the concepts of r- and K-selection apply equally well to micro-organisms and metazoans. He tested the hypothesis on four strains of Paramecium primaurelia, which represent groups of well-studied Protozoa. Here r-selection was applied to a population by restricting growth densities well below the limit set by the food supply, which if done frequently controls population size independent of density, thus permitting continuous growth at high densities without competition. The prediction that r-selection should reduce the saturation density was not proved. The results showed that r-selection favoured high values of both r_m and saturation density (K). Among the stains of P. primaurelia tested, better competitors had reduced saturation densities. Gili (1972), however, found that neither r_m nor K was related to competitive ability among strains of the Paramecium aurelia complex. When Luckinbill (1979) considered a range of ciliates of varying cell volumes and biomass, one aspect of the r-K selection hypothesis was apparent in Protozoa: the better competitors were larger species, with high individual population biomass. Large ciliates usually have lower growth rates than smaller species. The poorer competitors were small ciliate species, with lower overall biomass and high growth rates.

A different approach was adopted by Taylor (1978b), who tested the hypothesis with 11 species of bacterivorous ciliates in a small pond, comparing the measured or predicted innate rate of increase (r_m) of a species with its commonness in the field. The prediction was that those species with high values of r_m would be less common, as indicated by the number of samples in which they were found. Taylor used the index r/r, where r is the experimentally estimated r_m and r the r_m predicted by the equation of Fenchel (1968). A fast-growing species has a high r/r value, while slow-growing species have a low value for r/r. As predicted there was a negative relationship between r/r and commonness, the species with a high r/r value being less common than those with a low r/r value

(Figure 4.5). Paradoxically, the species designated as r-selected by r/r are mainly large species, but Taylor (1978b) argues that r/r is to some extent independent of size, and although size does enter into the calculation, an animal of any size or r_m can be assigned to the K-selected or r-selected category. Physiologically larger ciliates may be better adapted to withstand periods without food, and small species capable of more efficiently exploiting the small-scale spatial heterogeneity that occurs during periods of resource limitation.

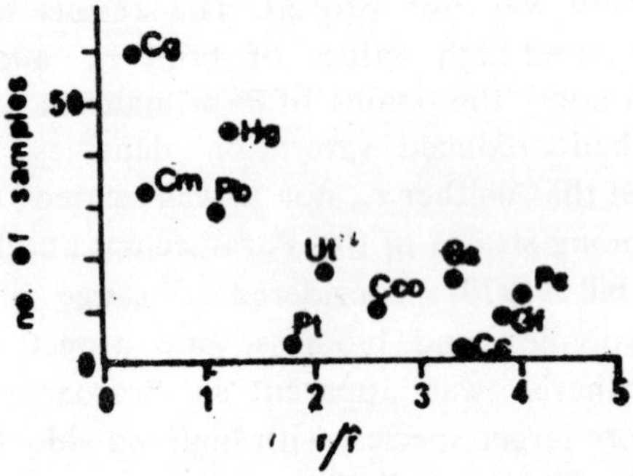

FIG. 4.5: The Relationship between r/r and Commonness among Ciliates, Cg—Cyclidium glaucoma, Cm—Cinetochilum margaritaceum, Cco—Colpidium colpoda. Cc—Colpidium campylum, Gf—Glaucoma frontata, Gs—Glaucoma scintillans, Hg—Halteria grandinella, Pc—Paramecium caudatum, Pb—Paramecium bursaria, Pt—Paramecium trichium, Ut—Urocentrum turbo.

Isolating populations in the laboratory and testing the r-K selection hypothesis can inevitably lead to anomalies. For example, using K as an index of competitive ability does not take interference competition into account (Gill, 1972), and neglects a whole array of environmental factors which impose an effect on feeding, growth and reproductive physiology of an organism, often over a short period in the natural environment. One factor which must not be overlooked is the very short life-cycle of protozoans compared to metazoans, so that the characteristics of a population can change quickly in terms of numerical very rightly points out, an organism pursuing a persistent or K-'strategy' in an unstable environment, may, if transferred to a stable one, succeed in the role of an r-'strategist'. An organism's position on the r-K continuum is influenced by the abiotic conditions imposed by the environment and the nature of the competition it encounters. Thus Protozoa cannot be recognised as being either r- or K-strategists; their position on the continuum is a function of an array of biotic and abiotic variables.

Interactions of Bacterivorous Protozoa with their Food Source

Microorganisms, particularly bacteria, play a role in the flow of carbon, nitrogen and phosphorus in nature, and additionally act as an energy source to a range of microbivorous organisms, including many species of Protozoa. The interactions of the Protozoa with their food source have been considered in a traditional predator—prey context. Early studies on the impact of Protozoa on their bacterial food source suggested that the relationship was not simply one of a secondary consumer exploiting a primary consumer; there seemed to be a stimulation of bacterial growth as a consequence of the presence and/or the grazing activity of the Protozoa. Since bacteria perform an important role in releasing nutrients from dead organic matter, and since their growth appeared to be modified by the presence of Protozoa, it follows that Protozoa may play an indirect role in nutrient recycling. Interest in decomposition processes and the recycling of nutrients in nature has grown in recent years, and evidence is now accumulating to suggest that Protozoa may play a direct, as well as an indirect role in the recycling of some nutrients, particularly phosphorus.

(i) The Impact of Protozoa on their Bacterial Food Source

Much of the early work on this aspect of protozoan ecology was directed towards soil-dwelling species, and in particular the impact on nitrogen-fixing bateria in the soil. Later, interest turned to the aquatic environment. All these studies indicate that usually the impact of protozoan grazing on bacteria is an increase in bacterial production and a hastening of the oxidation of decomposing organic matter in the soil or in water. Exactly how Protozoa achieve this effect is a matter of some debate. Other explanations include the secretion of growth-promoting substances by protozoa (Hervey and Greaves, 1941; Strasbrabova-Prokesova and Legner, 1966; Nikoljuk, 1969).

The level of nitrogen fixation by Azotobacter in the presence and absence of protozoan grazing has been widely investigated. It is a comparatively easy model to monitor in the laboratory. One of the earliest observations was made in 1909 by Russell and Hutchinson, who observed an increase in ammonia production when Protozoa were present in soil. Later Nasir (1923) and Culler and Bal (1926) showed that increased nitrogen fixation by Azotobacter occurred when the bacterium was grazed by ciliates or amoebae. In the presence of Azotobacter and other bacteria, nitrogen fixation in 50 ml of medium containing 0.25 g of mannitol amounted to 0.81-1.10 mg over 15 days, but in the presence of Colpidium, nitrogen fixed was 2.42 mg. Where Hartmannella grazed the bacteria, 1.69-2.21 mg of nitrogen was fixed under the same conditions in the 15-day period (Cutler and Bal, 1926). The ciliates and amoebae had the effect of reducing bacterial numbers, while enhancing nitrogen fixation by the remaining bacteria. The phenomenon was attributed to a greater efficiency in the use of the substrate where bacterial density was reduced by protozoan feeding activity.

Hervey and Greaves (1941) also found that the quantity of nitrogen fixed in the presence of a ciliate, in this case Colpoda, was greater than when Azotobacter was ungrazed. They considered a series of possible mechanisms which may have been responsible for the effect they observed. One possibility was the view that bacteria were maintained in physiological youth by protozoan grazing. Alternatively the Protozoa and bacteria had a type of mutualistic relationship, the bacteria serving as food, while the Protozoa excreted ammonia which maintained the alkalinity of the medium by neutralising the organic acids present. Neither theory sufficiently explained their results when tested, so they looked for other causes. The discovered that the addition of dead Protozoa still increased nitrogen fixation, which led them to the conclusion that Protozoa must secrete a growth-promoting substance. They suggested that it was probably a protein of an enzymatic nature, acting either as an absorbing agent rendering certain ions more accessible, or carrying vitamins or vitamin-like substances.

The concept of protozoans secreting a bacterial growth-promoting substance has been considered several times since it was first mooted by these authors (Hervey and Greaves, 1941). In aquatic situations ciliates enhanced the multiplication of bacteria in the presence of glucose during the initial 12-24 hours of incubation, during which time insignificant protozona multiplication occurred. The same effect on bacterial growth was achieved by adding liquid from which ciliates had been removed, suggesting the presence of some stimulatory substance. The stimulatory effect was destroyed by heating to boiling, which possibly indicates the denaturing of a protein (Straskrabova-Prokesova and Legner, 1966).

Nikoljuk (1969) found that amoebae and ciliates increased nitrogen fixation by bacteria in a range of soil types. Flagellates, however, which also consumed bacteria, had little effect on the level of nitrogen fixation by soil bacteria. The fact that some protozoan groups can stimulate increased nitrogen fixation and some do not to any great extent, may be a function of the varying impact of their grazing on the bacteria, but alternatively may be the result of the ability of some Protozoa to produce growth-promoting substances while others do not possess the capability to do so. Nikaljuk considered this possibility, and found that a growth-promoting substance was apparently produced, and in the case of amoebae was identified as indole-3-acetic acid (IAA) or heteroauxin, which is a plant growth substance. Azotobacter did not apparently produce IAA itself, but subsequently Brown and Walker (1970) found small amounts of IAA in aerated liquid cultures of Azotobacter chroococcum. However, the fact that the bacterium produces IAA itself does not preclude an enhancement of growth by additional IAA secreted by Protozoa. The production of this substance which stimulates plant growth is not unique to the Plant Kingdom, for human urine is a rich source of indole-3-acetic acid (Thimann, 1979).

The stimulatory effect of Protozoa on nitrogen-fixing bacteria has been shown to be temperature-dependent. Experiments with the ciliate Colpoda steini and Azotobacter chroococcum at a range of temperatures produced very varied results. At 28°C more nitrogen was fixed by the bacteria on their own, than in the presence of the ciliate, while at 25°C there was no significant difference in the amount of nitrogen fixed in the presence or absence of protozoan grazing. At 15°C, however, cultures with Colpoda produced twice as much nitrogen as pure bacterial cultures. At 5°C the ciliate was unable to survive (Darbyshire 1972). These results suggest that Colpoda may be near its thermal limit at 28°C; this is a high temperature and not one experienced for any length of time in temperate soils, although 28°C happened to be the optimum temperature for nitrogen fixation by Azotobacter. At 5°C the ciliate did not function at all, which suggests an optimum temperature for Colpoda around 15-20°C; at 15°C the ciliate was found to enhance nitrogen fixation. Darbyshire (1972) gave support to the growth substance theory with the proviso that it functioned only at some temperatures. In the soil a wide range of protozoan species occurs; each with different temperature tolerances. Thus at any temperature there will always be some species able to function efficiently and presumably maintain their intimate stimulatory effect on their bacterial food source.

The stimulatory action of Protozoa on bacterial production has recently been considered in the decomposer communities of aquatic environments. Fenchel (1977) showed a reduction in bacterial numbers to 50 per cent of the ungrazed density, when a mixed protozoan fauna was present. A single protozoan species succeeded in reducing bacterial density by 70 per cent. Much of the detritus in freshwater aquatic environments is plant-derived, and it has been suggested that the bacterial density on such detritus is closely related to the free surface area and that it is the availability of surface which may limit bacterial production (Fenchel and Jorgensen, 1977). In the presence of protozoan grazers, 20 per cent or less of the surface area of detritus is covered by bacteria, but in the absence of Protozoa all the available area on the detritus is colonised by bacteria. Another possibility suggested by Fenchel, and Jorgensen (1977) is that microturbulence created by the Protozoa may have a stimulatory impact on bacteria.

(ii) The Role of Protozoa in Nutrient Recycling

Nutrient regeneration can be defined as the release of soluble organic and inorganic nutrients, necessary for primary production, from dead organic matter. In terrestrial ecosystems decomposition processes brought about by microbial activity are confined to the soil and litter layers, whereas in aquatic environments they not only occur on and in the sediments of the benthic zone, which are analagous to soils, but in the planktonic communities of the water as well. Most of the earlier work on decomposition processes was focused on the soil habitat, where bacteria are one of the major elements resposible for the process of nutrient cycling. The relatively recent quantitative studies of decomposition and nutrient regeneration in aquatic environments indicate that bacteria are not the only important group of organisms involved. Indeed, much of the phosphorus and nitrogen incorporated into aquatic primary and secondary production may be regenerated by processes other than bacterial action (Johannes, 1968).

In aquatic habitats, many organisms, including zooplankton and Protozoa play a role as decomposers and nutrient regenerators, because as a consequence of their metabolising the organic molecules in their food, they excrete nutrients. In this way quantities of phosphorus in the organic orthophosphate form are released into the environment (Pomeroy *et al.*, 1963; Satomi and Pomeroy, 1965). Nitrogen is released in ammonia-free amino acids and other compounds (Johannes, 1968). Phosphorus in important in energy and nutrient cycles, and Protozoa have been shown to play an important role in the recycling of this element in marine waters. Johannes (1964) showed that per unit weight dissolved phosphorus excretion rates by marine animals increased with decreasing organism size. Since Protozoa are small and have a correspondingly high metabolic rate, they have a high rate of phosphorus excretion. The time taken to excrete the equivalent of the body content of phosphorus was less than one hour in the ciliate Euplotes (Johannes, 1965). This is extremely high when compared with zooplankton, which take 1.5-3 days to release the equivalent of their body conten of phosphorus (Pomeroy *et al.*, 1963).

Conflicting results were found in a tundra freshwater habitat by Barsdale *et al.*, (1974). While they were able to demonstrate that the transfer of phosphorus from detritus through bacteria to solution was enhanced by the presence of grazers, little of the phosphorus actually passed through the grazers in the process, because the turnover time of bacterial phosphorus due to grazing activity was about 24 hours, whereas the turnover of phosphorus in the bacterial biomass was only an hour or less. They were unable to explain the exact mechanism whereby phosphorus cycling occurred more rapidly when Protozoa were grazing bacteria than when bacteria were ungrazed, but suggested that protozoan removal of bacteria modified bacterial physiology and promoted a more rapid assimilation of the phosphorus contained in detritus and a more rapid circulation of phosphorus. Whatever the cause, conflicting though the evidence is from different types of ecosystem, it is clear that Protozoa play a valuable direct and/or indirect role in the cycling of phosphorus.

Carbon is one of the most important elements cycled in nature, since it forms one of the major components of living cells. Two areas of an organism's physiology are important in the carbon cycle, one being the intrinsic rate of inerease of an animal and the second being the rate of respiration, or the fermentation, of carbon. These two aspects determine the relationship between assimilated carbon and loss through respiration. Protozoa, by virtue of their rapid rates of growth and division and high assimilation efficiencies, will sustain a larger population on a given quantity of nutrients and energy than metazoans, and will also attain greater efficiency in the conservation of nutrients and energy. Hence the phosiological characteristics of Protozoa confer upon them a valuable role in the cycling of carbon (Stout, 1980). In experimental soil microcosms, where bacteria was grazed by amoebae, which in turn were predated by nematodes, the release of carbon through respiration was significantly higher than in ungrazed bacterial controls (Coleman *et al.* 1978). These workers found that in an experimental situation where fixed amounts of carbon were available, partitioning of carbon to biomass and respiration varied in relation to the complexity of the food chain and the assimilation efficiencies of the organisms involved.

Carbon is not usually a factor limiting primary production in terrestrial habitats, but may be so in aquatic environments, where the available CO_2 in the water may limit photosynthesis. However, it is normally the availability of nitrogen and phosphorus in any system which imposes constraints on the rate of primary production and bacterial decomposition. Where these essential elements are not available in the dead organic matter substrate in amounts sufficient to sustain bacterial growth, they are assimilated by the bacteria from the surrounding environment, so that in some circumstances the decomposers may be in direct competition with the primary producers for nitrogen and phosphorus. In the soil this phenomenon is known as nutrient immobilisation, because it is a process in which bacteria reduce the quantities of nutrients available to plants (Alexander, 1961). In aquatic environments bacteria can absorb phosphorus and nitrogen from the water and have the capability of storing phosphorus in excess of their immediate needs. It follows that high bacterial densities in the soil or sediments and waters of aquatic ecosystems is not necessarily indicative of a high turnover of nutrients. Studies on phosphorus transformations in artificial soil communities show that bacteria very quickly assimilate and retain most of the labile inorganic phosphorus available. Bacterial phosphorus is released only slowly when bacterial grazers are absent. When in the presence of bacterivorous amoebae or other protozoans, however, phosphorus is rapidly mineralised and returned as inorganic phosphorus to the pool (Cole *et al.*, 1978). Much faster nutrient recycling occurs extrinsic to the organic substrate between bacteria and bacterial grazers, and it is this part of the mineral recycling which controls the rate of decomposition processes and in addition the structure of the decomposer communities. The grazing activities of Protozoa on bacteria, their stimulatory effect on the rate of bacterial production, and their own intrinsically high rates of growth and reproduction all contribute to the valuable role played by these microfauna in the regeneration of nutrients.

Predators of Protozoa

Protozoa, along with other elements of the micro-and meiofauna, are subject to predation. There are considerable difficulties in identifying the predators of Protozoa by the usual procedures. Protozoa do not leave easily identifiable remains when consumed by a predator, so the techniques of gut and faecal analysis are of no value. Direct observation, usually in the laboratory, is the only reliable means of determining which predators readily exploit Protozoa as an energy source.

Most of the evidence in respect of predation on Protozoa pertains to aquatic environments. The medium lends itself more readily to laboratory investigations based on observation, whereas similar studies on soil are logistically more difficult. There is no doubt that Protozoa must be ingested by detritivorous soil organisms, since the protozoans will be intimately associated with the dead organic matter and its bacterial flora.

The question arises as to whether Protozoa are an essential source of energy to such organisms. Some experiments with the earthworm Eisenia foetida indicate that Protozoa are necessary for normal growth and are a valuable part of the normal diet (Miles, 1963). Among the Nematoda there are a variety of trophic types, including carnivores and bacterivores. It has long been suspected that carnivorous nematodes exploit Protozoa as part of their repertoire of prey. Recently this has been confirmed from studies on trophic interactions in experimental soil microcosms, where populations of Acanthamoeba polyphaga were reduced by the predation activities of the nematode Mesodiplogaster lheriteri (Anderson *et al.*, 1978).

In the aquatic environment crustacean elements in the zooplankton and benthic communities prey on Protozoa. Both filter-feeding Cladocera and Copepoda and raptorial-feeding Copepoda take a variety of Protozoa, and ciliates in particular. There is considerable evidence to show that Protozoa are a significant element in the diet of these small crustaceans. The raptorial benthic-feeding copepod Acanthocyclops bicuspidatus has been fed experimentally on algae, detritus and a range of ciliate species. The development times of A. bicuspidatus raised on protozoan diets were one-third to one-half the development times on algal or detritus diets (Strachan, 1980). Since the copepods maintain good production on Protozoa when compared to the other food sources available in the benthos, it is not unreasonable to suppose that copepods actively select these unicellular organisms as food in the natural environment.

Ciliates have been estimated as contributing as much as 20 per cent of the plankton biomass of inshore marine plankton communities, but until fairly recently have been overlooked as a potential energy source to other faunal elements in the zooplankton (Berk *et al.*, 1977). Porter (1973) noticed that ciliate numbers were reduced when planktonic copepods were present compared to controls where copepods were absent. Later Berk *et al.* (1977) showed that Eurytemora, a calanoid copepod, can significantly reduce small ciliate numbers, and further that the copepods derived nutritional benefit from them.

Ciliated Protozoa also appear to be a significant food source to zooplankton in freshwater environments. The filter-feeding cladoceran Daphnia magna extracts and consumes ciliates from the water filtered during feeding. Where large ciliates, such as Paramecium caudatum, are preyed on, only some of those entering the feeding current were brought inside the carapace, and into the food groove; most were broken up by the movement of the thoracic appendages, labrum and mandibles and ingested, but with the loss of some material. Fairly high levels of assimilation efficiency, of the order of 65-66 per cent, are achieved by D. magna on a ciliate diet (Porter *et al.*, 1979).

Field studies indicate that planktonic protozoa, especially flagellates and ciliates, are distributed in definite layers in the water column of lakes. In Dalnee Lake (Kamchatcha) maximal numbers in mid-summer occurred around 16-18 metres depth, which more or less coincided with the thermocline (Sorokin and Paveljeva, 1972). The thermocline is the area of interface between the unmixed lower hypolimnion and the upper epilimnion in stratified lakes. The maximal biomass of Protozoa developed in the period of algal death and the simultaneous development of large bacterial populations, at the end of the spring algal bloom. During the latter part of July the protozoan density decreased. Sorokin and Paveljeva (1972) attributed the decrease to predation by planktonic copepods and the rotifer Asplanchna priodonta. In July one-third of the primary food resources in the pelagic community of the lake are formed via bacterial mobilisation of dead organic matter. The most important role in the secondary trophic level is performed by Protozoa, which make a significant contribution to overall zooplankton production. The role played by Protozoa and their interactions with other organisms in Dalnee Lake is shown in Fig. 4.6.

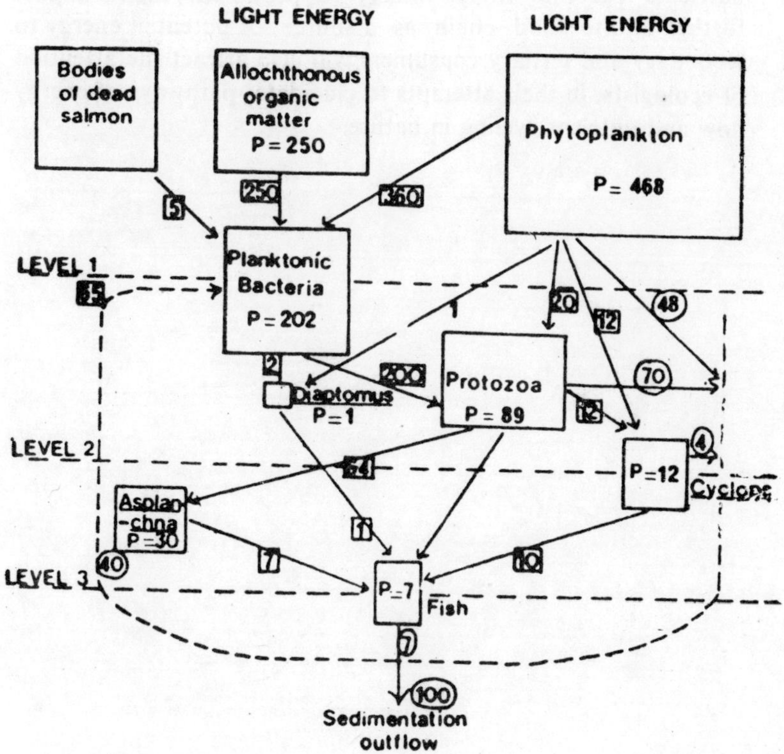

FIG. 4.6: The role of Protozoa in the plankton of a Lake Community. The figures show production for 30 days in June. P=production, the numbers in squares are rations and the numbers in circles are non-assimilated food.

Source: Sorokin and Paveljeva (1972), with the permission of Dr. W. Junk, Publishers.

The value of Protozoa as a food source to other organisms in the food chains of soil and aquatic benthic or planktonic communities is an area in which we have only limited information at present. Undoubtedly, as the realisation that Protozoa play a role in decomposition processes and the recycling of nutrients becomes more widely appreciated, their impact further up the food chain as a source of potential energy to secondary and tertiary consumers will also attract the attention of ecologists, in their attempts to elucidate pathways of energy flow and nutrient cycling in nature.

5

Ecological Adaptations in Protozoans

Despite the fact that Protozoa are ubiquitous, they are frequently overlooked in faunal surveys of ecosystems, but as we have seen in the preceding chapter they play a valuable role in the functioning of many ecosystems. Much of their success in colonising so varied an array of habitats is due to their physiological and behavioural adaptability. Through time various assemblages of species have evolved physiologically to meet the variety of physical, chemical and biological conditions imposed by a range of ecological niches, many of which are inhospitable or hazardous. One of the advantages of a single-celled organisation is that it is often more readily able to undergo evolutionary adaptation and change than a multicellular level of organisation. The widespread ability of Protozoa to encyst, which is a characteristic shared by a range of distinct taxonomic groups in the Protista, has proved invaluable to those species which have successfully moved from aquatic environments to terrestrial and semi-terrestrial habitats.

The position of many natural protozoan populations in the saprovore food web has resulted in their ready colonisation of artifical systems set up by man to deal with waste water and sewage. Sewage treatment plants are essentially simple artificially manipulated saprovore systems aimed at purifying the waste organic materials of urban industrialised society. The protozoan communities of sewage treatment plants, particularly ciliates, perform a voluable role in the production of good quality clean effluents (Curds, 1973, 1975; Curds and Cockburn, 1970a, b).

The small size, rapid growth rates and the ease with which some species can be cultured in the laboratory, together with the tolerance of many Protozoa to polluted conditions, has led to the consideration of these organisms as indicator species, and as species for bioassay in pollution studies. Perhaps more importantly the impact of bacterivore Protozoa on the population dynamics of bacteria and in turn on decomposition processes, suggests a valuable role in polluted environments where they may enhance the decomposition of polluting agents of organic origin.

Ecology and Adaptation in the Natural Environment

Protozoa have colonised a wide spectrum of aquatic and terrestrial habitats worldwide from the arctic to the equatorial zones. Many of the studies on soil Protozoa have focused on testate species (Heal, 1962, 1964; Couteaux, 1975, 1976; Lousier, 1976), which tend to give a biased view of these protozoan communities. In some soils testate Protozoa are a dominant group, as they are in the antarctic soils of Signy Island where ciliates are few and naked amocbae totally absent. Here testates and flagellates constitute the major components of the soil Protozoa (Smith, 1973). In some climates, however, ciliates, flagellates and amocbae may be abundant (Stout, 1962; Rogerson and Berger, 1981 b). In mineral and beech litter soils Stout (1962) found ciliates to be more numerous than rhizopods, and interestingly the community tended to be dominated by only a few species—in mineral soils by about five species and in beech soils by about ten. Rogerson and Berger (1981 b) found amoebae and flagellates to be the most numerous in garden soil, followed by ciliates, with one recurring helizoan species and an occasional testate amoeba during the period from November to April.

Soil species face variations in the water content of their environment; desiccation is particularly pronounced in the summer months in some climates, and in other soils water becomes locked up in ice during the winter months. Although we talk of terrestrial or soildwelling Protozoa, they are not strictly terrestrial because their normal function depends on the presence of the surface films of water around soil particles, or on vegetation such as *Sphagnum*, in which they live. Soil Protozoa are essentially aquatic organisms living in a hazardous environment, particularly where climatic extremes prevail. Physiological adaptation is an essential prerequisite to successful population function among soil Protozoa. Where desiccation of the upper soil levels occurs, the Protozoa can migrate downwards to a limited extent providing there is sufficient oxygen and water, and the soil structure allows it, or alternatively they can encyst. Similarly encystment can be resorted to when soil waters freeze. The evidence suggests that encystment is a readily available refuge when environmental conditions become adverse to the continuance of normal physiological functioning. The process of encystment is described in detail in the following section.

Semi-terrestrial salt marsh ecosystems are usually extermely productive. Webb (1956) found a varied protozoan fauna present in the saltings of the Dee estuary in Cheshire, with ciliates occurring as the most abundant group. The Protozoa of the mudflats, however, showed a less varied community. She was unable to find any correlation between the most common ciliates and gross fluctuations in temperature, salinity or hydrogen ion concentration or in relation to seasonal or tidal changes, which suggests that species commonly found in this habitat are physiologically robust and well adapted to coping with the varying conditions encountered in a salt marsh. Lee and Muller (1973) found that Foraminifera of a ralt marsh were highly productive, indeed as productive as other groups, including nematodes. The high productivity of forminiferans in this ecosystem indicates an important role in the overall energy flow of the community and the cycling of nutrients.

In aquatic ecosystems there are two habitats available to Protozoa, the pelagic and benthic zones. The benthic zone is usually the richest in terms of organic matter and microflora, hence it is usually more heavily populated by protozoans than the open waters. However, flagellates and ciliates commonly occur in the plankton and have been reported as a substantial part of the pelagic community in one freshwater lake (Sorokin and Paveljeva, 1972). In the North Atlantic amoebae have been found to be common in the subsurface of the planktonic zone (Davis *et al.*, 1978). Further detailed studies of pelagic protozoan communities and population dynamics are sorely needed. The indications are that in some aquatic environments planktonic Protozoa may be a significant element in the overall planktonic food web.

Studies devoted to the benthic community as a whole, or focusing solely on the protozoan element in the benthos, have been carried out in aquatic habitats since the beginning of the century. Birge and Juday (1911) noted active representatives of a number of protozoan groups under anaerobic conditions in the hypolimnion of Lake Mendota. More detailed studies of fresh water benthic protozoan communities followed (*e.g.* Wang, 1928; Moore, 1939; Cole, 1955; Webb, 1961; Goulder, 1971, 1974) and later intensive studies of the benthos of marine and brackish waters appeared (*e.g.* Fenchel and Jansson, 1966; Muus, 1967; Fenchel, 1967, 1968).

In the majority of lakes, except those which are very shallow, stratification of the waters occurs in a yearly cycle. The result is that for parts of the year the benthic zone is subject to low temperatures and depleted oxygen in temperate lakes. The more eutrophic the lake the greater the depletion of oxygen, so that in highly eutrophic lakes complete deoxygenation of the profundus occurs. Such conditions are not conducive to normal physiological function in obligative aerobic organisms, and consequently changes occur in the abundance and species composition of the benthic protozoan community. Webb (1961) recorded 128 species of Protozoa in a small eutrophic lake, Esthwaite Water, in the English Lake District. The abundance of Protozoa, the majority of which were ciliates, was related to the availability of oxygen in the benthic zone. Later surveys on the same lake (Goulder, 1974) showed that large ciliate species such as *Frontonia leucas*, *Loxodes magnus*, *L. striatus*, *Spirostomum minus*, *S. teres*, *Stentor coeruleus* and *S. polymorphus* generally reached their maxima when the bottom of the lake was oxygenated, but the majority dis-appeared in the summer months when anoxia prevailed. Other species, which are clearly anaerobic or facultatively so, such as *Caenomorpha medusula*, reached their maxima under deoxygenated conditions. Faced with the adverse summer regimen the normal aerobic protozoan fauna has two avenues of avoidance available—either migration upwards into the water column or laterally to the benthic littoral zone into oxygenated conditions, or encystment.

Detailed studies of the benthic zone and the overlying water column have revealed seasonal migration patterns by many of the species commonly found in the fine silty muds of the lake bottom, when oxygen becomes limited (Finlay, 1981; Bark,1981). Dense populations of the large ciliate species listed above, of the order of 3,633±944 cm $^{-2}$, are typical from October to May. *Loxodes* sp. and *Spirostomum* sp. dominate the community (Finlay, 1981). During summer stratification most of these ciliates migrate upwards into the overlying water, the most densely populated zone coinciding fairly closely with the 0.3-1.0 mg l^{-1} oxygen isopleths. The vertical migrations involve considerable distances for such small organisms; at certain times the majority of protozoans are some 8-9 metres from the sediment surface. While some species migrate, others such as *Paramecium* disappear completely. A group of small ciliates, less than 150 μm in length, remain on the lake sediments developing large populations. These include Metopus, Caenomorpha medusula, Brachionella spiralis and Saprodinium dentatum (Bark, 1981). Thus the benthic habitat is vacated by aerobic forms and a community of sulphide ciliates, similar to those in marine communities described by Fenchel *et al.* (1977), develops on the sediment surface and in the overlying waters (Finlay, 1981). A few of the species which undertake seasonal migrations are also found in low densities on the sediments during anoxia, which has led to the supposition that they may be facultative anaerobes, able to switch from one form of metabolism to another in response to prevalent environmental conditions.

The sediments of freshwater lakes are usually fine mud or silt so that penetration to any depth by Protozoa is impossible. Most species occur in the upper centimetre, although some individuals do penetrate down to 3-4 cm in the sediments (Goulder, 1971; Bryant and Laybourn, 1972/3). In stratified lakes the anoxic community is restricted to the sediment surface and its occurrence is seasonally determined by the pattern of stratification.

The migratory behavioural patterns of the normally benthic Protozoa of stratified lakes are particularly interesting. Unlike soil-dwelling species in which encystment is the main means of coping with adverse environmental regimes, freshwater ciliates are able to adopt a behavioural means of avoidance. It is probable that encystment is a less common occurrence in the aquatic environment than in the soil habitat, although undoubtedly species must resort to encystment if they are unable to migrate, respire anaerobically or if their temperature or pH limits are exceeded. As an adaptation to soil colonisation, soil-dwelling protozoans have come to employ encystment more readily, encysting and excysting over short periods as conditions temporarily deteriorate and improve (Rogerson and Berger, 1981b).

In the marine ecosystem the benthic communities of the profundal areas are still largely a mystery, although foraminiferans have been found at depths up to 4,000 metres (Coull *et al.*, 1977). It is the littoral and sublittoral zones which for obvious reasons have been studied in detail. Fenchel (1969) tentatively classified the microfaunal benthic communities of the marine inshore environment into estuarine and sand microbiocenosis, the sublittoral and microbiocenosis and the sulphuretum. The estuarine sand microbiocenosis comprises the microfaunal communities of sandy sediments in fjords, lagoons and sheltered shallow bays. These areas often have high levels of organic material resulting in reduced conditions near to the sediment surface. The sublittoral sand microbiocenosis is found in fine to medium sands possessing a low organic matter content, which are permanently submerged. The sulphureta are biotopes dominated by the sulphur cycle.

The estuarine sand biocenosis has a rich ciliate fauna which is largely confined to the surface oxygenated sediments. A wide spectrum of trophic types are found including species exploiting diatoms, blue-green algae, bacteria, other microfauna and purple and white sulphur bacteria near the sediment surface. The lower reduced sediments contain ciliates exploiting sulphur bacteria under anaerobic conditions. The dominant process in marine sediments is sulphate reduction; the sulphide produced may be bound to iron in the anoxic zones producing a characteristic black colouration. The specialised sulphur ciliates living in these zones are members of the Trichostomatida, Heterotrichida and Odontostomatida. The latter order contains only anaerobic species which lack cytochrome oxidase and mitochondria (Fenchel, 1978). The sublittoral sand microbiocenosis harbours a large number of characteristic ciliates, many of which feed exclusively on diatoms. Bacterivores and predators are also found in this group. Sulphureta, which are common along the shores of many fjords, lagoons and inner bay waters, represent parts of the coastal and marine ecosystem dominated by the sulphur cycle. Thriving populations of ciliates have established themselves in this habitat; Fenchel (1969) recorded some 45 species of ciliate among the sulphur bacteria in the sulpuretum Niva Bay.

Encystment

Encystment is a widespread phenomenon among taxonomically diverse protozoan groups. Undoubtedly the ability to encyst plays a valuable role in the life strategies of Protozoa, and since it is a characteristic so widely shared it probably developed early in their evolutionary history. The cyst allows a species to withstand adverse conditions in a state which is in some respects analogous to the conditions of hibernation and diapause in higher animals. Additionally the cyst may act as a dispersal stage for many species, particularly those inhabiting terrestrial and transient aquatic environments. A significant number of Protozoa are dispersed in the air or externally and internally by animals. Cysts have been observed in aerial and animal samples, which when cultured have given rise to viable organisms (Corliss and Esser, 1974). The resting cyst plays a valuable role in the lifecycle of those species having the capability of entering such a physiological quiescent state.

During the process of encystment, a series of very radical changes occurs in the cell. There is considerable dedifferentiation, both morphologically and physiologically. The most obvious morphological changes are the loss of the organellar systems of cilia and flagella in most ciliates and flagellates, the eventual loss of the contractile vacuoles and food vacuoles and changes in the structure and distribution of ultrastructural components in the cell.

The nuclei and nucleoli undergo marked changes associated with considerable activity, but the changes appear variable among Protozoa. In Stylonychia mytilus the macronuclei fuse to form a single macronucleus, and the chromatin becomes organised into discrete round or oval-shaped bodies. Two types of nucleoli develop in the macronucleus during encystment: a centrally-placed larger nucleolus and a smaller peripherally-located diffuse nucleolus which disappears in the encysted cell (Walker *et al.*, 1975). Fusion of macronuclei also occurs in Oxytricha, but here the nucleoli become homogeneous (Grimes, 1973). In both cases the micronuclei resemble those in the vegetative form. The macronucleus of Blepharisma stoltei undergoes shortening, adopting a horse-shoe shape and in the later stages of encystment develops a vacuole (Repak, 1968). In Acanthamoeba sp. buds form from the nucleolus, which apparently do not contain nucleolar material. The buds are later incorporated into autolysosomes (Griffiths, 1970). A recent study on Acanthamoeba astronyxis found that protrusions developed from the nucleoli during encystment which appeared to pinch off as nucleolus-like bodies. One or two of these bodies per nucleus were observed (Lasman, 1982). The exact reasons for all of these variable transformations are unknown.

Mitrochondria also undergo changes in the structure and distribution in the cell. In Stylonychia mytilus the mitochondria form tightly packed aggregates which late in encystment form incomplete bands peripherally situated in the cyst. The organelles appear irregular in shape but Walker *et al.*, (1975) attributed this appearance to the close packing. Grimes (1973) also noted that mitochondria aggregated in bands in Oxytricha; here however they were situated around, but distinct from, the central macronucleus. Changes in mitochondria of Acanthamoeba have been reported by Vickerman (1960), who found intra-cristal inclusions. During encystment in Acanthamoeba astronyxis circles of rough endoplasmic reticulum appear enclosing protoplasmic inclusions, especially mitochondria. Simultaneously the mitochondriaelongate and constrict (Lasman, 1982). Many of these changes are probably associated with changes in the secretory and synthetic activities of the cell during the encystment process. Autolysosomes have been reported during encystment and are prominent throughout the process (de Duve and Wattiaux, 1966; Bowers and Korn, 1969; Holt, 1972; Geimes, 1973). These vacuoles contain various cellular organelles and migrate towards the cell surface where they discharge their contents. The discharged material may become trapped in the outer layer of the cyst wall, the exocyst.

During encystment there are changes in gross cell composition. A reduction in cell volume occurs, which can be considerable. In Colpoda steini, for example, a 60 per cent reduction in cell volume associated with a substantial decrease in cell water content has been reported (Tibbs and Marshall, 1970). As one might expect, contractile vacuole activity is considerable during encystment, increasing in the later stages to achieve cytoplasmic shrinkage. Total cell protein decreases in Colpoda steini to 30 per cent of the original amount, and of the remaining protein some 15 per cent is locked up in the cyst wall (Tibbs and Marshall, 1970). Similar reductions in both volume and protein content occur in Acanthamoeba where a very large proportion of cell protein, around 33 per cent, is incorporated into the cyst wall (Neff *et al.*, 1964; Neff and Neff, 1969; Rudick and Weisman, 1973). The change in protein content can be followed through growth deceleration and the encystment process. As the organism passes from the exponentially growing phase to the stationary phase and the advent of starvation, which induces encystment in a number of species including Acanthemoeba sp., there is a doubling in cell protein levels, followed by a progressive decline during encystment. During the transition to the stationary phase there is extensive proliferation of cytoplasmic membranes, which are subsequently broken down as encystment progresses (Pauls and Thompson, 1978).

The cyst wall is composed of a number of layers which usually range from two to four depending on the species. Indeed there appears to be considerable variation in the structure of the cyst wall among Protozoa. Walker *et al.* (1975) described four layers in the cyst wall of Stylonychia: an outer exocyst which appears either amorphous or lamellar; a mesocyst which is fibrous with a lamellar configuration; an amorphous endocyst; and a layer which envelops the cytoplasm composed of compacted membranous material thrown into folds. In Oxytricha fallax the exocyst, mesocyst and endocyst are similar to those of Stylonychia, but the innermost layer is composed of granular or coarse lamellar material (Grimes, 1973). Nassula ornata has two membranes in its cyst wall, a faceted lamellate exocyst and a thin endocyst (Beers, 1966). The exocyst of Acanthamoeba castellanii is described as having two layers, an outer amorphous layer overlying a fibrillar layer which may contain some amorphous material, cell debris and glycogen granules (Bowers and Korn, 1969). There may, however, be some variation in the number of layers described by researchers. The exocyst described by Bowers and Korn (1969) could be designated as two layers. The endocysts of several species of amoebae, including Acanthamoeba, have been shown to contain a cellulose component (Neff *et al.*, 1964; Rastogi *et al.*, 1971). The proportion of cellulose, protein, lipids and carbohydrates in the cyst wall varies from one species of protozoan to another.

While encysted, a protozoan is in a quiescent physiological state. There is no growth or reproduction and the maintenance of living material is continued at a fraction of the respiratory energy expenditure necessary in the vegetative form. Obviously the encystment process itself requires increased metabolic activity which is reflected in increased oxygen uptake, but at the termination of the encystment process the respiratory rate of the cysts in Hartmanella was not measurable with Warburg respirometry (Griffiths and Hughes, 1969). Even with the more sensitive cartesian diver microrespirometer the respiratory function of protozoan cysts is not measureable (Laybourn, 1976c). A conservation of energy during the encysted phase, which in extreme cases can last several years, is essential if an organism is to have energy in hand for the energetically expensive process of excystment and commencement of normal locomotion, feeding and growth.

The factors responsible for inducing encystment are probably far more complex than may be apparent from laboratory studies, where often one factor such as depletion of food supply as in Acanthamoeba castellanii, Oxytricha fallax, Nassula ornata, Blepharisma stoltei and Colpoda steini (Hashimoto, 1962; Repak, 1968; Tibbs and Marshall, 1970; Paul and Thompson, 1978), or high temperature as in Nassula ornata collected from low temperature conditions (Raikov,

1962), induces encystment. Indeed, the fact that different strains of one species, Nassula ornata, have been found to respond to both lack of food and adverse temperatures by encysting, suggests that both factors operate, or that physiologically distinct strains from one habitat adapted to a particular set of environmental conditions may respond in a different manner to other strains adapted to other regimens. Environmental factors other than temperature have been implicated as inducing encysment and include low oxygen concentration, elevated or reduced pH, increased salt concentration and the accumulation of metabolites (Corliss and Esser, 1974).

We have very little information on why and how encystment occurs under natural conditions, but there are one or two studies which give some indication of the occurrence of encystment in nature and the conditions and factors which may be responsible. Some species of soil and freshwater testate amoebae encyst during the winter months, obviously in response to low temperatures and possibly also reduced food supply, while in the summer months they are active (Schonborn, 1962; Heal, 1964). In the case of zoochlorellae bearing testate species, for example, Hyalosphenia papilio, encystment in the later part of the year occurs before temperatures fall. In this case it is probably reduced light intensity which induces a reduction in the number of active individuals in the population in Sphagnum swards (Heal, 1964). Rogerson and Berger (1981 b) observed rapid encystment of the protozoan community of a soil when temperatures declined below 0°C. Rapid excystment occurred when temperature increased, providing there was sufficient soil moisture. Some intertidal species appear to encyst and excyst in response to a tidal rhythm. Strombidium occulatum swims freely at high tide in the water, sinking to the bottom and encysting in intertidal pools at low tide (Faure-Fremiet, 1948).

There appears to be some variability in the ability of the cysts of different species to withstand desiccation. Some freshwater species, for example Euplotes and Didinium, possess cysts which cannot withstand drying (Beers, 1937; Garnjobst, 1937), whereas the cysts of some soil species can tolerate long periods under dry conditions. Dawson and Hewitt (1931) revived Colpoda cucullus after five years of dry storage. The aquisition of resistance to extremes of desiccation in soil species is obviously an evolutionary adaptation to an environment of a more hazardous character than the aquatic ecosystem from which they originated. The extremes of conditions which can be withstood by the cysts of species from different environments is poorly researched, leaving one rather interesting aspect of protozoan ecology and life-history strategies to be elucidated.

The process of excystment requires a stimulus, either replenishment of food supply, suitable temperature or some other factor or combination of factors. The process can be readily induced in the laboratory in some species. Didinium nasutum excysts in response to the presence of bacteria. The presence of its prey, Paramecium, alone is insufficient (Butzel and Horowitz, 1965). The fact that Paramecium does not cause excystment in Didinium casts an interesting light on the relationship of this very specialised predator to its prey. Didinium cannot encyst without first feeding, so that if it excysts and fails to locate prey, it perishes. However, dense bacteria are likely to attract bacterivorous ciliates including Paramecium. Nassula ornata rapidly excysts within 2-3.5 hours in the presence of bacteria. Beers (1966) has described the series of events occurring during excystment in Nassula in detail. The first sign of cyclosis is the appearance of the contractile vacuole, which swells and having reached a size of about 30 μm, empties very slowly. The process is repeated several times during cyclosis. After 2-2.5 hours a slight elevation develops at the posterior end of the cyst and over the surface of the elevation the exocyst becomes thinner so that its facets are obliterated. It appears that pressure is exerted from within the cyst, probably as a result of the entrance of water, initially into the cytoplasm and later into the contractile vacuole. The contractile vacuole then ceases to evacuate and swells to a diameter of 65 μm. The exocyst reptures and some of the cytoplasm protrudes abruptly. The form of the repture is always a slit, when may be as narrow as 15 µm, so that the animal is severely constricted as it passes out, apparently by cytoplasmic streaming even though at this stage the animal is fully ciliated.

In other protozoans the cyst wall simply ruptures, presumably as a result of water uptake or in other cases as a result of enzymatic break-down of the cyst membranes. Often, as in the case of Didinium nasutum, a combination of two processes is involved in excystment. In this species a large excystment vacuole forms which causes the rupture and forces off of the exocyst and mesocyst. The vacuole then contracts and a small Didinium can be seen swimming inside the endocyst which gradually increases in size and dissolves, thus liberating the enclosed animal (Beers, 1945).

Ultrastructural changes during excystment have been observed in Oxytricha fallax and include a complex sequence of micronuclear division. The excysting cell contains dividing, non-dividing and seemingly reabsorbing micronuclei. The function of this micronuclear behaviour is unclear, but in any case, by the time the newly emerged animal is ready to undergo its first division the normal complement of two micronuclei and two macronuclei is established (Grimes, 1973). Dedifferentiation of the ciliature in ciliates occurs during excystment. The first sign in Oxytricha is the appearance of a few kinetosomes at the cell surface, which proliferate and become primordia. Then sequentially the cirral primordia, marginal cirral primordia and dorsal bristle primordia develop (Hasimoto, 1963; Grimes, 1973). These structures develop de novo, there being no visible markers for the sites of primordia formation. Very shortly after excystment is initiated, the sub-surface sheet of microtubules is formed. By the time the animal emerges from the cyst it has a fully formed ciliature.

The ability to encyst has allowed the Protozoa a wider range of habitat colonisation. Soils in which Protozoa are a successful group would in most cases not be a viable environment without the refuge of the encysted state during periods of adversity when desiccation or when freezing of interstitial waters occurs. Terrestrial environments are subject to more rapid and often more severe physical and chemical change than aquatic environments. Encystment for a protozoan is an essential aspect of the life-cycle evolved to withstand periods of adversity. Higher organisms have evolved more sophisticated means of coping with seasonal environmental extremes, by hibernation, diapause or as in the case of many annual invertebrate species, by overwintering the population in the egg or pupal stage. All these mechanisms are designed to cut physiological function to the minimum, so that the organism passes through an adverse phase when temperatures are low or when food is limited, with the smallest possible expenditure of energy.

Protozoa in Sewage Treatment

In addition to colonising a wide array of habitats in natural ecosystems, Protozoa have established themselves in habitats created by man for the treatment of used water. Sewage and used water are treated in a number of ways to produce relatively clean effluents which can be discharged into the environment. In Britain two processes are widely used, the activated-sludge plant and the percolating-filter bed. The former involves the activation of the sludge so that the oxygen incorporated allows bacterial multiplication which leads to oxidation of the organic material in the sewage. The percolating-filter involves thickling the used water, after primary settlement, over a deep bed of small rocks or an artifical medium. The bed has outlets at the bottom through which the purified effluent passes. The medium, usually rocks, is colonised by bacteria which perform the same function as they do in activated-sludge. Activated-sludge plants represent a somewhat unstable environment for colonisation, but nevertheless adaptable elements of the micro- and meiofauna, including Protozoa, Nematoda and Rotifera, have established themselves in this food-rich environment. Percolating-filters are physically stable habitats and consequently have a greater diversity of animals including copepods, annelid worms, insects and other Metazoa in addition to the groups found in activated-sludge.

Protozoa are abundant in activated-sludge and percolating-filter beds. Densities of 50,000 cells ml^{-1} have been recorded in the mixed liquor of activated-sludge, which represents about 5 per cent of the dry weight of suspended solids in the liquor (Curds, 1973). Achieving quantitative estimates of the densities of organisms in percolating filters is extremely difficult, because the organisms are distributed over the surface of the rocks or plastic media, but observations indicate that Protozoa are abundant.

Essentially the sewage treatment plant is analogous to the polluted natural environment, and the organisms which colonise such habitats must be able to withstand the conditions which are characteristic of pollution. In natural ecosystems severe organic pollution reduces the fauna to all but the most robust species, which often flourish in polluted conditions because of reduced interspecific competition and the rich food supply in terms of dead organic matter and the associated microflora. Protozoa are frequently abundant under such a regimen, hence their successful colonisation of used water processing plants.

Protozoa associated with sewage treatment have been considered by various workers (Ardern and Locket, 1928; Baines *et al.*, 1953) but the most comprehensive study of Protozoa and their role in sewage treatment processes has been carried out by Curds and his co-workers. A survey of 47 percolating filters and 52 activated-sludge plants in the British Isles conducted by Curds and Cockburn (1970a) showed that ciliates were the dominant group, followed by Rhizopoda and Phytomastigophorea in descending order of abundance. The Actinopodea were represented by one species, Actinophrys sol, which was infrequent in its distribution. In total 67 species of ciliate were identified in activated-sludge plants, while the effluent of percolating-filter beds revealed 53 species. The film scraped from the surface of the medium on the top of the bed showed less species diversity than the effluents, which suggests a succession of species through the depth of the filter bed (Curds, 1975).

In a detailed study of the Protozoa occurring in the activated-sludge process, Curds and Cockburn (1970a, b) found a correlation between the composition of the protozoan community resident in a plant and the quality of the effluent produced. Plants which delivered effluents of high quality with low BOD (Biochemical Oxygen Demand) and low suspended solids, contained a wide variety of ciliated Protozoa in high densities, whereas those plants which produced poor quality effluents contained no ciliates and only a few other Protozoa in low numbers. Laboratory investigations using small-scale activated-sludge plants have proved the important role of Protozoa in waste-water treatment. Under Protozoa-free conditions six replicate plants produced very poor quality turbid effluents, the turbidity being a function of high bacterial numbers and suspended solids. BODs were in the range 53-70 mg l^{-1} and suspended solids ranged between 86 and 118 mg l^{-1}. The addition of cultures of Protozoa produced a very dramatic improvement in the quality of the effluents discharged. After only a few days the BOD dropped to 7-24 mg l^{-1} and suspended solids to 26-34 mg l^{-1}. The improved BOD was attributable to a decrease in viable bacterial numbers from $106\text{-}160 \times 10^{6}$ ml^{-1} to 1.9×10^{6} ml^{-1} (Curds *et al.*, 1968).

There are several possible explanations for the effect on effluent quality brought about by Protozoa. Bacterivorous ciliates are capable of consuming extremely high numbers of bacteria and voracious grazing on dispersed bacterial growth is probably the major mechanism—indeed Curds believes this to be the case. Some ciliates possess the ability to flocculate suspended particulate matter and bacteria (Watson, 1945; Curds, 1963). This process has also been implicated in the production of good effluents, but Curds (1975) does not consider the process to be of any great import; however, he suggests that amoebae may have the ability to ingest the flocculated bacteria.

The ciliates found in sewage treatment plants can be categorised into three distinct groups: first, free-swimming species which either move in the liquor of activated-sludge or in the surface microbial film of the media in filter beds; secondly, crawling ciliates which crawl on the surface of the floc, and lastly, stalked forms which attach to the floc or media (Curds, 1973). During maturation of activated-sludge there is a succession of protozoan species. Buswell and Long (1923) found that holotrichs were the first to appear, giving way to other forms. Stalked species, such as Carchesium and Vorticella, appeared after several weeks aeration. More recently Curds (1966) carried out a detailed study of the succession of colonising species in activated-sludge using small-scale experimental plants. Colourless flagellate species, Oikomonas socialis, Peranema trichophorum, Heteronema acus and Bodo were the first to appear. Soon after, ciliates of the free-swimming category appeared, including Paramecium caudatum, Oxytricha fallax, Oxytricha ludibunda and Uronema nigrans. These were replaced by crawling hypotrichs, such as Aspidisca castata and Aspidisca lyneus. When the sludge reached maturity after being aerated for 3-4 weeks peritrichs appeared (Figure 5.1). Curds (1966) considers nutrition to be the major controlling factor in the succession. The pioneering flagellates are either holozoic, saprozoic or saprophytic and all are species normally associated with waters of high organic content. The colonising ciliates then follow successions of bacteria.

The Role of Protozoa in Polluted Ecosystems

In industrialised societies pollution is an ever-present problem. The need for rapid and easy tests for indicating the effects of pollution on the flora and fauna and the ecological balance in the environment receiving pollutants, is obvious. Protozoa are near the bottom of the food chain and their small size and rapid reproduction renders the sublethal impact of pollutants more readily discernible than in larger animals with longer,

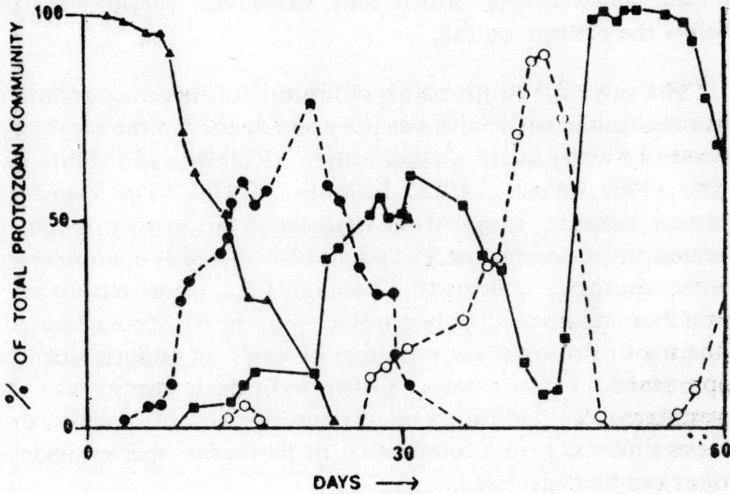

FIG. 5.1: The Succession of Ciliates in Developing Activated-sludge.

often complex life-cycles. Consequently ecologists have experimented with Protozoa as potential indicator species. An additional aspect is the potential of Protozoa in aiding the breakdown and neutralisation of pollutants. The preceding section dealt with the valuable role of Protozoa in aiding the purification of sewage in artificial habitats, and it follows that as far as organic material is concerned the same function may be performed by protozoans in natural aquatic ecosystems. In oil pollution Protozoa may also play a role in degradation processes.

Aquatic habitats are more prone to pollution than terrestrial environments because water is so widely used in industrial processes, and the effluents of both industry and urban developments are deposited in moving waters, *i.e.*, streams and rivers, estuaries or the sea, for dispersal. Most of these ecosystems can cope with a degree of pollution and usually more severe pollution is reflected in a change in the normal faunal and floral make-up of the community. Severe pollution may create a total antoxic zone which may extend for varying lengths below the effluent outfall.

The value of Protozoa as indicators of organic pollution was recognised early in this century and applied in the saprobity, system of water quality classification (Kolkwitz and Marsson, 1908, 1909; Wetzel, 1928; Leibmann, 1936). The saprobity system spans a range from unpolluted to grossly polluted (xenosaprobic—unpolluted, oligosaprobic—scarcely contaminated β-mesosaprobic—moderately contaminated, α-mesosaprobic—heavily contaminated, polysaorobic—very heavily contaminated), The use of Protozoa as indicators of levels of pollution can be approached in two ways. Either individual species can be investigated as indicators of various degrees of pollution, or species diversity in a community or particular species associations can be considered.

At the species level Sladecek (1964, 1969) produced an indicative value of a species in relation to a degree of saprobity. Thirty-five species of free-moving ciliates were given an index ranging from 1-5 based on their occurrence in nature under particular degrees of saprobity. This approach suggests that some species lend themselves well as indicators, occurring commonly under a particular set of saprobic conditions. Species with a wide distribution through saprobity have obviously less value.

At a more specific level, the effects of isolated chemicals commonly found in industrial effluents have been tested on Protozoa for acute and chronic toxicity evaluation. Apostol (1973) exposed Paramecium caudatum to sodium chloride, potassium carbonate, ammonium nitrate, magnesium sulphate, ferrous orthophosphate and lead acetate in concentrations from 1.0 mg l^{-1} up to 10,000 mg l^{-1} in acute and chronic tests. The chronic testing involved carrying out experiments in nutrient media to determine any effect on vitality and reproduction, whereas in acute tests the ciliate was simply exposed to the chemicals to evaluate the impact on survival time and any cell damage. Survival time in acute tests decreased with increased toxicant concentration, lead acetate being the most toxic with survival times ranging from 0.1 minute at 10,000 mg l^{-1} to 500 minutes at 500 mg l^{-1}. Magnesium sulphate proved the least toxic. Under the chronic testing regime reproductive rate was depressed, usually in relation to increasing toxicant concentration. Bacterial density was not monitored, and there is the possibility that bacterial reproduction was suppressed by a toxicant which would then diminish protozoan food supply, leading in turn to a reduction in reproductive rate in Paramecium.

A similar approach, testing the impact of toxicants on an aspect of protozoan physiology as a bioassay technique, was adopted by Slabbert and Morgan (1982). They used the respiratory response of Tetrahymena pyriformis to poisons such as ammonia-N, cyanide, PCP, phenol and mercury and elicited a response within 10 minutes to toxicant concentrations in the range 0.5-5.0 mg l^{-1}. In some cases intoxication caused a decrease in oxygen uptake, in others a marked increase. The latter response was attributed to a state of hypersensitivity indicating that any increase in toxicant concentration would prove lethal. Since the majority of toxicants were detectable at concentrations very much lower than the levels specified for industrial effluents, this technique may be applied successfully for the detection and control of toxic industrial discharges, although at present it is not sufficiently sensitive to detect toxicants at the lower levels permissible in drinking water. However, there is the possibility that other species of ciliates may prove more sensitive to poison and would be more suitable for testing drinking water. The use of Protozoa in bioassay techniques on specific poisons common in industrial effluents is a relatively new area of applied protozoology, but the results indicate that ciliates may prove useful organisms in an important area of ecological monitoring.

In a broader approach, communities of Protozoa may be used for the biological assessment of water quality. Particular associations of organisms are assumed to indicate different degrees of decomposition in organic pollution. In much the same manner that a succession of Protozoa can be observed in the maturation of activated-sludge, successions of associated species can be observed in the process of decomposition in natural waters. Bick (1973) followed the dynamics of protozoan populations in the breakdown of organic matter in a model freshwater ecosystem. The decomposition process takes some time, usually about 7-10 days, to reach the stage where a distinct succession in species and high numbers of ciliates are observed. Successive dominants were Glaucoma scintillans, Cyclidium, Halteria, Colpidium, Coleps hirtus, Chilodonella cucullulus, Stylonychia putrina, Paramecium caudatum, Litonotus lamella, Acineta and Microthorax. Most of the species are bacterivorous, although one or two carnivores, Litonotus and Acineta, are present exploiting the other Protozoa as food. These organisms are characteristic of the first three weeks wherein a high intensity of decomposition is established with attendant environmental conditions of oxygen depletion and the presence of ammonia. Bick (1973) terms this a heterotrophic phase. In the later stages when conditions begin to improve an autotrophic phase develops, harnessing the nutrients released by decomposition, when only small numbers of ciliates are present. The large ciliate community is replaced by Euglena, Chilomonas and algal species. The succession of species is regulated by biological and environmental factors such as the availability of food, oxygen, the toxic products of decay, competition and predator prey relations.

The implications of these successions is that protozoan communities clearly play a role in the purification of polluted waters in a manner similar to that which occurs in sewage plants, but also that the community structure is indicative of particular conditions. The real problem comes in correlating the species assemblages present in reasonably accurate terms to the degree of saprobity, especially when one bears in mind that the various stages of saprobity merge into each other. Dresscher and Mark (1976) have devised a simple formula to overcome the problem of the transition from one phase to another and the importance of various groups of Protozoa and algae. The saprobic quotient (x) is calculated as follows:

$$x = \frac{C + _3D - B - _3A}{A + B + C + D}$$

The letters indicate the number of forms of each group of organisms. A is ciliates marking polysaprobity, B corresponds to Euglenophyta (Euglenida) indicating α-mesosaprobity. C corresponds to Chlorococcales and Diatomeae indicating β-mesosaprobity, and finally D related to Peridiaeae, Chysophyceae (Chrysomoadida) and Conjugatae indicating oligosaprobity. The limits for the formula are −3 (polysaprobic) to +3 (oligosaprobic). Zero is therefore at the borderline between the condition α-mesosaprobic/β-mesosaprobic and β-mesosaprobic/α-mesosaprobic. The method has the limitation that it cannot be applied when there is a predominance of one species.

In order to monitor an environment quickly for its level of saprobity, it is necessary to sample the water, sediments and macrophytes for the presence of Protozoa. Muddy sediments are time-consuming to analyse and stones and macrophytes are difficult to sample quantitatively. Ideally one requires a rapid means of determining which species are present and their relative abundance. The use of artificial substrates and their colonisation has been investigated by Cairns and his co-workers (Yongue and Cairns, 1978; Cairns *et al.*, 1979), who have shown that the communities which occur on polyurethane foam substrates reflect the natural community. Recently, Henebry and Cairns (1980) have used protozoan communities on polyurethane foam sudstrates to monitor pollution in a stream subject to enrichment by ammonia, nitrates and phosphates from industrial discharges. Their results suggest that protozoan communities on artifical substrates are an effective means of assessing the degree of pollution, and compare favourably with the macroinvertebrates frequently used as biological indicators. Protozoa on artificial substrates have the advantage of being easy to collect and examine, whereas sampling and collecting macro-invertebrates may be expensive in terms of time.

The evidence to date suggests a valuable role for Protozoa in environmental monitoring aquatic ecosystems. This is a function of the ecological adaptability of many species and the fact that their physiology responds rapidly to various toxicants, so that sublethal effects are manifested quickly. As bioassay techniques employing Protozoa become more refined there is no doubt that they will become more widely employed in this sphere.

The problem of oil Pollution occurs not only in aquatic ecosystems but also in terrestrial ecosystems where oil is extracted, and on salt marshes where estuasine spillages from oil terminals may accumulate by tidal action. There are a variety of bacterial species capable of degarding crude oil. The question arises as to whether there are Protozoa which exploit these bacteria and whether such Protozoa function normally under conditions of oil pollution. Obviously oil-degrading bacteria may be stimulated by protozoan grazing in the same way that bacteria involved in natural decomposition processes are stimulated. Experiments have indicated that crude oil has no adverse effects on freshwater and soil Protozoa, although there is a tendency to slightly larger cells in the presence of oil. The species studied maintained good growth on a diet of oil-degrading bacteria. During feeding some oil is ingested but no deleterious effects are apparent.

High latitude ecosystems are now becoming exploited for oil. Cold temperatures are not conducive to high biological activity and hence to the degradation of oil pollution in such climates. Rogerson and Berger investigated the impact of oil pollution on soil in Ontario during the winter months in an attempt to mirror the conditions of more northerly climes. They found that amoebae were the most important group of soil Protozoa, followed in abundance by flagellates and ciliates. Oil had no effect on the protozoan community which exploited the high densities of bacteria associated with the oil. When temperatures dropped to sub-zero the Protozoa encysted, but rapidly excysted when temperatures rose. Indeed during the winter months many species encysted and excysted repeatedly in response to temperature changes although growth and reproduction were minimal during the short active phases. These reoccurring protozoan populations may be expected to have a significant impact on the microflora. Rogerson and Berger suggest that under arctic conditions, where the trophic activity of many terrestrial invertebrates is severely limited during the winter, Protozoa may be extremely important in the functioning of the soil, particularly when subject to pollution.

In aquatic environments oil pollution is normally treated with dispersants. Frequently it is the dispersants rather than the oil which causes death or danage to flora and macrofauna. The impact on Protozoa is poorly researched, although the impact of Corexit 9527 on a range of species has been investigated. The tolerance of each species varied with increasing concentrations. In each case, however, growth rate was constant up to a critical threshold at which point death occurred. Rapid detoxification of cultures occurred in the laboratory, possibly due to evaporation and/or microbial degradation, which suggests that a more rapid recovery from conditions toxic to the most sensitive protozoans would prevail in the field. Since Protozoa can be expected to be relatively unaffected by dispersants in nature it follows that their grazing activities and possible stimulatory impact on oil-degarding bacteria would function even when dispersants are employed.

6

Ecology of Bryozoans

Colonies grow by proliferation of new zooids and extrazooidal tissues. Colony form is determined by the position of new zooids with respect to those formed earlier, and their shape, orientation, and rate of addition in each portion of the colony. Patterns of zooidal addition determine the range of colony morphologies that potentially and actually exist, and therefore constitute both a phylogenetic and architectural constraint on colony form. In this chapter we describe the ways bryozoans grow into different colony morphologies.

Encrusting Growth

New zooids are added to encrusting colonies along part or all of the colony margin. In the first case, growth results in runner-like colonies if the zones of active growth are narrow and divide at regular intervals. If these zones are broader, irregular in width, and divide at irregular intervals and widths, then irregularly lobate sheets result. In contrast, growth around the entire margin produces more radially symmetrical forms. Frontal budding may also increase the thickness (height) of lobate or radially symmetrical encrusting sheets.

Runners

Runners grow as uniserial or narrow multiserial branches that increase by bifurcation at branch tips or by lateral branching at places along the branch. Runners extend rapidly and, depending on the rate of branching, may explore growth in several directions along the substratum.

The geometric constraints and consequences of highly patterned runner growth may be seen in two species of Jurassic Stomatopora that are formed of dichotomous branches. Within fossil colonies the bifurcation angle averages 155° for the first but diminishes to and stabilizes around 75° by the fourth and subsequent bifurcations. Gardiner and Taylor ran a series of simulations of colony growth in which the angle of bifurcation remained constant across all generations, or was systematically increased or decreased with successive generations. They found that any constant angle of bifurcation from 1° to 180° results in branch convergence after growth of three sets of links of equal length, and that progressively increasing the angle of bifurcation only accelerates this convergence. However, progressively decreasing the angle of bifurcation delays branch convergence in proportion to the rate of decrease. Branch convergence in fossils apparently caused termination of one of the branches and loss of any descendant branches that might have derived from it. Thus, delay in convergence due to astogenetic decrease in bifurcation angle should have increased colony size, zooid number, and reproductive potential.

Single and Multilayered Sheets

Encrusting sheets add new zooids along continuous peripheral margins. Commonly, new zooids are added at similar rates around the entire periphery of the colony, resulting in radially spreading buttons or sheets, although variable rates of growth or the presence of obstructions result in colonies with irregular outlines. Stronger polarity in growth rates of encrusting sheets results in the belt-shaped colonies such as develop in Membranipora membranacea on fronds of the alga Laminaria hyperborea (Ryland and Stebbing 1971). Most of these colonies acquire a growth orintation that is roughly proximal down the algal stipe before they reach 3.5 cm diameter, and 80 per cent of all colonies—some up to more than a metre in length—grow as broad belts straight down the frond.

Encrusting sheets growing on limited substrata may overgrow themselves where actively growing edges meet, or where a vigorous edge encounters an older, senescent portion of the colony. In this manner multilayered colonies of either stenolaemates or cneflostomes may form through self-overgrowth. Self-overgrowth also occurs in stenolaemates in which zooids, either by differential growth or, more typically, as a result of damage to their neighbours, extend above the adjacent apertures. The overtropping zooids establish a new multizooidal budding zone that extends laterally over the surrounding surface. This occurred widely in Paleozoic trepostomes where local areas either became sensecent or were damaged (Bigey 1981).

Multilayered growth by various forms of frontal budding is common among Cenozoic and Kecent cheilostomes. The best studied example is Schizoporella unicornis (Banta 1971, 1972), in which the body cavities above frontal calcified shelves of the parential zooids balloon upwardly as the intrazooidal origins of buds. Vertical walls of new zooids are upward extensions of vertical walls of parental zooids, and interserial communication organs between adjacent "vertical" lineal series are established where exterior cuticle between series is locally dissolved. Lidgard also found that multilayered growth can result from coalescent multiserial budding and from nonlineal multiserial budding. In Celleporaria magnifica, among others chaotic, frontally budded regions include a mixture of the latter two types of budding.

Multilayered growth is typical of mound-like and nodular sheet-like to massive colonies. 'It has developed commonly in species that grow on limited substrata, such as pagurid-inhabited gastropod shells and other small substrata that develop as rolling stones (cheilostomes—Rider and Enrico

1979). It is also the predominant mode of growth among abundant species on rock walls and rests.

Erect Growth

Erect colonies have much more complex form than do encrusting colonies, and their growth is correspondingly more difficult to describe. To aid description, and to provide possible insight into the geometric and adaptive bases of different erect growth forms, there has been increasing use of models and simulations of growth of these organisms. Results have shown that very complex three-dimensional forms can be generated by specifying only a few parameters.

Unilaminate and Bilaminate Sheets

Among the erect growth forms, unilaminate and bilaminate sheets seem to exhibit the least regularity in their development. They grow essentially as encrusting sheets that have escaped from the substratum.

The growing edge of many encrusting species may lift off the substratum to form erect unilaminate sheets (Cheetham 1971, Jackson & Buss 1975). This may occur when the growing edge contacts an obstruction or the limit of the substratum, or for no apparent reason. Bilaminate sheets may be produced either haphazardly or as part of a regular, closely coordinated process. For example, where the cheilostome "Hippodiplosia" insculpta encrusts algae and gorgonian stems, the advancing edges of a single colony may meet where they have wrapped entirely around the substratum (Nielsen 1981). This results in the growing edges lifting up off the substratum and continuing growth back-to-back at a more or less coordinated pace, producing an erect bilaminate sheet. Nielsen noted that growing edges of different colonies of "H" insculpta can also lift off of the substratum where they meet and encrust one another to produce bilaminate sheets, and he noted at least one interspecific example formed by a colony of "H." insculpta and another of Thalamoborella californica.

Growth of bilaminate sheets in stenolaemates begins where a ridge-like wall forms within the colony. This wall becomes a median wall within a confluent multizooidal budding zone (Boardman 1983). Two oppositely facing budding zones are centered on the median wall and generate new zooids back-to-back as the median wall advances. The advance of the growing edges of both sides of the bilaminate sheet are therefore exactly coordinated by the common median wall.

Arborescent Bilaminate (Adeoniform) Colonies

Colonies composed of multiserial bilaminate branches, with zooids opening on both flattened sides, and that are continuously and rigidly calcified from base of attachment to branch tips, are termed adeoniform, following the usage originally intended by Brown (1952). Growth of colonies of most adeoniform species follows regular and precise species-specific rules of growth that fluctuate little as the colonies grow. The principal results are significantly reduced overlap and variation in spacing between the growing tips of branches, which further increases both the potential growth and surface area of colonies.

Growth of adeoniform colonies involves four processes:

(a) continuous branch elongation by budding of new zooids;

(b) branch widening, which results from multiplication of zooid rows;

(c) branch multiplication by division of the growing tip; and

(d) branch thickening proximal to the growing tip, which does not extend the colony into space but contributes to colony strength.

Bifurcation angle and mean length of branch links (segments between successive bifurcations) are constant throughout the cheilostome colonies that have been studied, although link length varies substantially and unevenly in some Paleozoic adeoniform species. In most species, two branches extending from a single bifurcation grow at unequal rates, whether they be from the first bifurcation above the substratum or one formed in later growth stages. Moreover, fully formed branch links extending from a single bifurcation are unequal in length, the inequality forming a characteristic ratio for each species examined. Each branch path consists of the sum of the links from the proximal end of a colony to a specific branch tip,

$$L_t=(1/G_t)(NaA+NbB)$$

and the total path length within a colony is

$$T_t=naA+nbB$$

where L_t is mean path length; G_t is number of branch tips; Na and Nb are the summed number of links of length A and B, respectively, in each path; T_t is total path length; and na and nb are the total number of links of length A and B within a colony. The relative rates of increase in L_t and T_t reflect colony business and can be important in adaptive morphology, for the former measure is proportional to height and the latter to surface area. The wide variation in T_t:L_t with growth of Cenozoic adeoniform cheilostome species reflects differences in height and bussiness among them. The most highly significant differences are between bifurcation angle and link length. Interspecific differences in link length seem to be major contributors to the differences in T_t:L_t, for when the two values are divided by mean link length (l) the differences in ratio largely disappear.

Variation in angle of bifurcation as measured in the plane of the parent branch affects the overall shape of a colony. Relatively low angles of bifurcation cause an increase in colony height, while higher angles cause an increase in colony radius. Overall colony radius is maximum in models where the angle of bifurcation is 90°. In some Paleocene, Oligocene, and Recent adeoniform species, the angle of bifurcation is greater than 60°, even though this results in branch loss at later growth stages when the lowest branches approach the substratum. There has apparently been selection for such high angles of bifurcation during the Cenozoic, for there has been a statistically significant ($P < 0.002$) trend through time towards the 90° bifurcation angle (Cheetham 1986a).

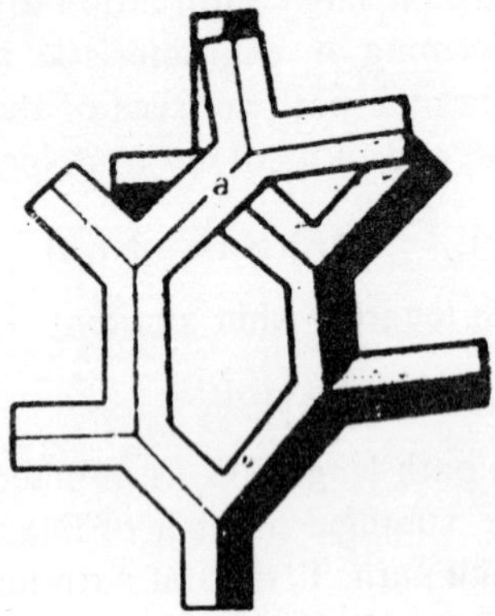

FIG. 6.1: Crossover of converging adeoniform branches (point a) generated by twist at bifurcations proximal to the point of convergence. (After Cheetham & Hayek 1983, courtesy of the Paleontological Society.)

Bifurcations in adeoniform cheilostomes—and in almost all the bilaminate arborescent stenolaemates—occur in the plane along which the two sheets of zooids join. If the plane remained perfectly flat, the median branches would converge into the same space after the third bifurcation and before the fourth could develop, just as in uniformly bifurcated models of encrusting runners. However, the newly developed branches at each bifurcation are twisted with respect to the ancestral branch. As a result, the first pair of converging branches cross over past one another and continue independent growth (Fig. 6.1). The crossover distance in living species averages about five times the initial branch thickness and increases with both the angle of twist and the angle of bifurcation; but it is only the twisting at bifurcations that makes the difference between three-dimensional bushes and perforated sheets.

Bifurcation angle, link lengths, and angle of twist do not change with growth, so that branches eventually begin to converge and to interfere with one another. This sets a limit to growth stage and size of colony, which is approximately 3(A+B) and is defined as Growth Stage 3. Growth Stages 1 and 2 are (A+B) and 2(A+B), respectively.

Increase in the twist angle causes an increase in the colony's radius perpendicular to the plane of its stem with respect to the radius in the plane of its stem. Regardless of the degree of twist, as colonies grow they continue to increase this ratio, which reaches a maximum at or near Growth Stage 3. The overall colony radius, and also the average nearest-neighbour distance of branch tips at Growth Stage 3, are at maximum if the angle of branch bifurcation is 90° and the angle of twist is 50°. Thus the Cenozoic increase in angle of branch bifurcation has also contributed towards increasing the overall colony radius and the average distance between branch tips at Growth Stage 3 in modern species.

As a colony grows, the average nearest-neighbour distance among branch tips begins a steady decline at the first branch crossover, but there is an even greater decline in minimum distance, which results in increased variability of spacing with growth. The space between branch tips is not evenly distributed at Growth Stage 3, the coefficient of variation reaching 50—60 per cent in both real and simulated colonies, so that it is not all available to accommodate branch growth. This high variation in spacing of branch tips at later growth stages contributes to their interference. The slightly lower level of variation in recent species over that in Paleocene and Oligocene species ($\chi^2=4.0$, $P<0.05$) means that there has been a phylogenetic decrease in branch interference.

Decrease in variability of nearest-neighbour spacing of branch tips is strongly polyphyletic. It appears to be due to whatever advantage accompanies reduced irregularity in spacing between branch tips, which is in turn the result of adhering more closely to a regular growth pattern.

Arborescent growth allows exponential increase in surface area. In adeoniform colonies, branch width and thickness of the growing tips are such that, by Growth Stage 3, surface area is about two times that of an encrusting sheet with the same diameter, although before Growth Stage 2 an encrusting sheet of equal diameter has the greater surface area. Massive hemispherical colonies and adeoniform colonies at Growth Stage 3 that have equal diameter also have approximately equal surface areas, but the arborescent colonies require only about 2—3 per cent the skeletal volume of massive hemispherical colonies, and thus have, presumably, considerable savings in investment in growth.

Arborescent Unilamiante Colonies

Unilaminate arborescent bryozoans are composed of closely spaced, narrow branches that are usually arrayed in well-developed flat or curved sheets. Each branch and intervening space is generally 1 mm or less wide, collectively forming a slotted or perforated sheet. Colonies tend to have similar shapes and apparently follow grossly similar rules of growth as do other colonial suspension feeders (gorgonians, hydrozoans, and graptolites) with narrow branches closely arrayed in planar sheets (McKinney 1981b). Such planar arrangements of branches are unique among colonies in their very narrow variance in which and spacing of branches. Although transport of materials along branches and problems of structural support may also contribute to their organization, the restriction of this type branching system to colonial suspension feeders suggests that filtration is a critical selective factor in its development.

The simplest unilaminate colony form is a wedge-shaped, regularly expanding, planar fan that may be either pinnate or dendroid. Form and texture of pinnate colonies are determined by

(a) distance between the origins of lateral branches on each side of the main branch;

(b) apposition or alternation of lateral branches:

(c) angle between lateral branches and main branch; and

(d) whether growth is continuous, forming an expanding fan, or is discontinuous, perhaps forming a roughly parallel-sided fan where growth of proximal branches has stopped.

Form and texture of dendroid colonies, whose branches increase by bifurcation, are determined by

(a) the arc or angle subtended by the growing margin as measured from the colony origin,

(b) centering of each branch between its two neighbours, and

(c) the critical lateral distances between three neighbouring branches at which the central branch will bifurcate.

Relationships between length of the colony from origin to growing edge, length of the growing edge, and surface area in uniformly expanding planar fans and cones are quite regular. The number of branches has a direct linear relationship to length of the growing edge. Therefore branch bifurcations, which are the means of increase in branch number, accumulate distally in direct proportion to length of the growing margin and to the spacing of branches within the fan.

Branch origins in pinnate colonies are highly predictable; they are concentrated along the main branch, and lateral branches have small probabilities of giving rise to descendants. In contrast, dendroid unilaminate colonies tend to have bifurcations scattered across the growing margin. The increased number of branches at their growing margins results in a distal decrease in the probability of bifurcation of any given branch per increment of increase in fan length.

Most unilaminate arborescent colonies do not grow as simply expanding planar fans or cones. More common are fans and cones that expand distally at variable or exponential rates and complexly folded sheets. In some of these more complexly shaped colonies, branch bifurcations are quite obviously concentrated in certain regions, such as along the left margin of the fenestellid colony illustrated. In this colony, the margin adjacent to the prolific bifurcations has spread and recurved over some of its more proximal parts. This example gives insight into the geometry of branching in regularly spiralled colonies that evolved from fan-shaped ancestors, which has occurred in the Paleozoic fenestrates (Archimedes) and in modern cheilostomes (Bugula).

Spiralled Bugula turrita colonies can be subdivided into a series of dendroid planar fans that originate from bifurcations along the axial helical margin of the colony. Serial surfaces cut through the heavily calcified axial margins of Archimedes skeletons demonstrate that its colonies are organized in the same general way.

The spiralled form can be modelled solely by establishment and maintenance of a bifurcating, helical margin in addition to the growth of the radiating simple planar fans (McKinney & Raup 1982). The configuration of the bifurcating, helical margin is determined by

(a) the radius of the helical path,

(b) the rate of climb of the helix (CLIMB),

(c) the radial angle between bifurcations along the helical path, and

(d) the angle between the axis of the helix and the branches that diverge from it (ANGLE).

The diverging branches each serve as the beginning of a planar branch system that is bounded by similar proximal and distal branch systems.

The several parameters of the models of spiralled colonies may be varied to produce a hypothetical morphospace within which the observed morphologies of species of Archimedes and Bugula fall. The observed morphologies in Archimedes and

Bugula show a narrow range in ANGLE but a substantial inter-specific variation in CLIMB. The interesting questions of why the actual morphologies are distributed as they are have yet to be answered by more than preliminary hypotheses.

Arborescent unilaminate colonies vary much more in overall form than do adeoniform colonies. This greater variability appears to result from more freedom in location of branch bifurcations and lengths than occurs in the adeoniform colonies. Some species of unilaminates (cones and spirals) have species-specific colony forms, while others are more indeterminate, with the ultimate form and size dependent in large part on unpredictable clustering of bifurcations and branch growth.

Among the colonial suspension feeders that form planar sheets, the active feeders (bryozoans and graptolites) are unilaminate and have a very narrow range of branch sizes and spacings, whereas branch organization, sizes, and spacing are more variable in the passive feeders (gorgonians and hydrozoans). Although colony geometry in all arborescent unilaminate bryozoans can be described in terms of branch spacing and position of bifurcations that add new branches, the stimuli for these attributes are at present unknown. Nevertheless, widespread similarities in branch size and spacing, and the range of colony shapes and sizes among distantly related arborescent unilaminate bryozoans suggest a common function.

Free-living Growth

Free-living stenolaemates were common from the Ordovician to Permian and grew by lateral growth over fine-grained sediments or by developing a concentric, potentially rolling habit. Concentric self-overgrowth and potentially rolling habit has also developed in post-Paleozoic cheilostomes. However, the most common free-living bryozoans are cap-shaped (lunulitiform) cheilostomes. Larval metamorphosis typically occurs on grains that are smaller than the eventual adult colony size.

Nonlineal budding has been characteristic of the free-living cheilostomes (Tavener-Smith & Williams 1972, Hakansson 1973, Lidgard 1985b). New zooids are partitioned from the ring of colonial space by rapid local proliferation of secreting epithelium situated low on the distal sides of the existing zooids. The newly forming skeleton eventually curves up to contact and fuse with the frontal cuticular wall (Fig. 6.2). Basal, lateral, and transverse walls all lack cuticle, are interior, and their growth divides new zooidal spaces from the undivided colonial space around the periphery, which is a nonlineal budding zone. When the colony grows beyond its initial substratum, the basal wall becomes separated from the basal cuticle by a space lined with epithelium on both surfaces. The space may be undivided, or various combinations of transverse and longitudinal cuticles form at some distance behind the growing edge and connect the basal skeletal wall and the basal cuticle. This anchors the basal cuticle to the skeleton. As the free-living colony grows peripherally, the more proximal skeleton continuous to thicken

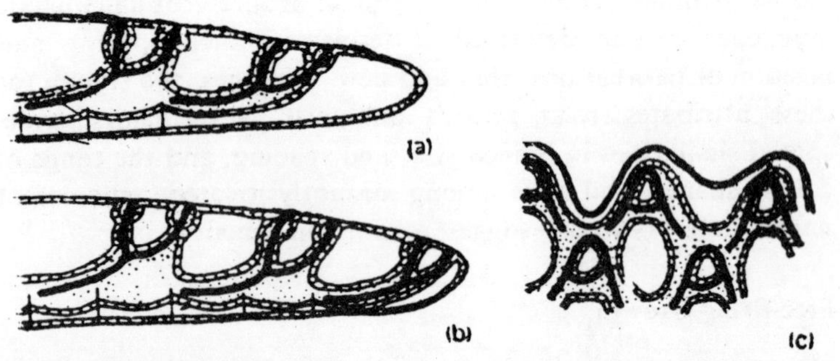

FIG. 6.2: Idealized sections through a free-living cupulariid cheilostome. (a) Longitudinal section cutting colonial space from which an autozooid will be partitioned at colony margin on right, (b) longitudinal section cutting skeletally defined vibraculum at colony margin on right, (c) tangential section through autozooids, vibracula, and marginal colonial space. Brick pattern represents epithelium, heavy black lines represent primary skeleton, stippled pattern represents secondary skeleton. (After Hakansson 1973, courtesy of Academic Press).

on its underside, often completely engulfing the original substratum.

Rooted Growth

Specialized adaptations for living rooted in sediments have evolved in several groups of cheilostomes and ctenostomes. These include a variety of forms that resemble various erect and free-living colonies. These rooted taxa, however, have turgid, heterozooidal or extrazooidal, cuticle-encased tubes that extend from the proximal or basal regions of the colonies and attach to grains of sediment. Other rooted forms are specially adapted for life on sediments, including numerous top-shaped, globular, conical, or stellate colonies of only a few millimetres width and length.

Calcified parts of top-shaped zoaria are not attached to any substratum, making interpretation of earliest astogeny inferential even for modern colonies, since early colony growth has not been directly observed. Rather than producing a single ancestrula, larval metamorphosis apparently produces a group of primary zooids, most of which are autozooids but with one, or perhaps more, differently shaped, rootlet-bearing kenozooid(s) (Fig. 6.3). Orifices of the autozooids are oriented towards the substratum, from which growth of the rootlet lifts the colony. Budding adds new zooids on the side of the colony opposite the kenozooidal apex. Thus the "distal" growth direction of the colony is opposite the orificial ends of autozooids, which contrasts with all other bryozoans.

In top-shaped colonies of the families Conescharellinidae and Orbituliporidae, the frontal wall has expanded at the expense of all the other wall regions. Marginal pores around the flattened hexagonal, oriffice-bearing portion of the wall (Fig. 6.3) are inferred to be the loci from which frontal buds originate. Autozooids formed in successive series are progressively larger in most top-shaped colonies of these families, so that a region of repetition of zooidal morphology never develops. Additional gradients in zooidal morphology may

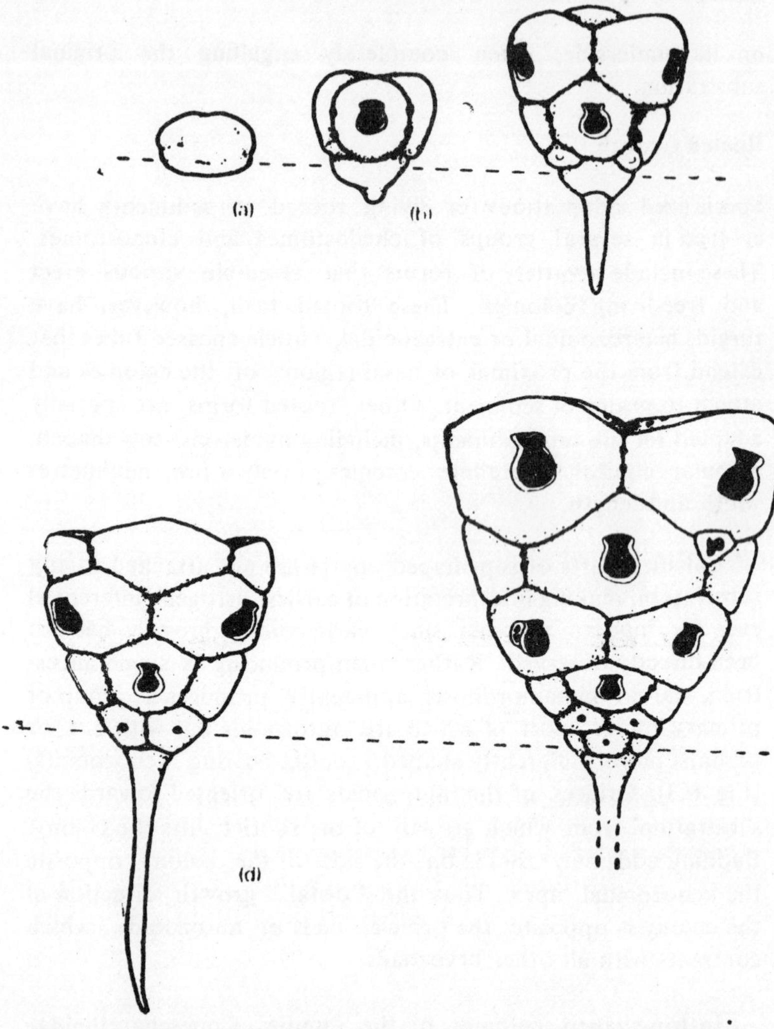

FIG. 6.3: Astogeny of idealized, rooted, coneschorellinid cheilostome. Larval metamorphosis (a) produces several primary zooids, (b) including one or more rootlet-bearing kenozooids (bottom). Continued growth of the colony, (c)–(e) adds progressively larger autozooids at the top of the colony and additional rootlet-bearing kenozooids at the proximal end. Note that orifices of autozooids are orientated proximally with respect to the direction in which they are added to the colony. (After Cook & Lagaaij 1976, courtesy of the British Museum of Natural History).

develop simultaneously with, or subsequently to, the initial zone by frontal budding from exposed frontal walls of zooids that are proximal to the distal edge of the colony.

Top-shaped colonies of genera in the family Mamilloporidae develop by more typical distal and disto-lateral budding. In these the distal growth direction of the colonies and orientation of autozooids correspond. Differences in orientation and growth of these colonies compared with the Coneschareilinidae and Orbituliporidae are probably related to opposite polarity of ancestrular complexes that are developed at completion of metamorphosis.

7

Growth Forms as Adaptive Strategies

Bryozoans have always grown in many shapes, but these can be reduced to a few basic forms. Bryozoan colonies differ in the ways zooids are arranged relative to each other (unserial versus multiserial) and in their principal directions of growth, which can be horizontal (encrusting) or vertical (erect). In uniserial colonies, one zooid typically buds another end-on, thereby forming single or loosely branching chains (for convenience we include biserial forms here). In multiserial colonies, zooids bud others laterally as well as distally to form more-or-less continuous surfaces of zooids instead of chains.

There are five basic growth forms that include the great majority of Recent and fossil Bryozoa characteristic of hard substrata:

(a) uniserial (and biserial) encrusting,

(b) multiserial encrusting,

(c) massive multiserial encrusting,

(d) multiserial erect, and

(e) uniserial (and biserial) erect.

Two additional multiserial growth forms, highly specialized for life on sediments, are free-living and minute rooted species.

In this chapter we compare the distributions of the five most common growth forms among four major extant groups of marine Bryozoa. Next we develop a simple model to describe the apparent costs and benefits of these five forms relative to risks of mortality from different causes. These deductions lead to testable predictions about the zooidal characteristics and environmental distributions of species with different growth forms. These predictions are generally sustained for both living and fossil bryozoans, which suggests that risks of mortality have selected for different growth forms in much the same ways throughout the Phanerozoic. However, the relative intensity of different factors has apparently changed greatly, as evidenced by massive temporal shifts in environmental distributions of the different growth forms.

Frequencies of Growth Forms among Major Taxa

A compilation was made of all bryozoan species reported in 47 ecological surveys and taxonomic monographs for the Atlantic basin from the Arctic Ocean to South Africa, and for the western Mediterranean. Effort was made to avoid duplication of species using available synonymies. This produced 3,300 records for a total of 1,430 species including 1,146 cheilostomes (419 anascans, 44 cribrimorphs, 683 ascophorans), 99 ctenostomes, and 185 cyclostomes. Growth form was determined from the papers or from original taxonomic descriptions; depth and substrata were recorded when available (Jackson, unpublished).

Distributions of the five forms for all taxa combined, more than half the species are multiserial encrusting, and the next most abundant are multiserial erect. The least common form is uniserial encrusting. Multiserial species are about four times commoner than uniserial, and encrusting 1.7 times commoner than erect.

The distribution of the five forms is strikingly different between major taxa. Ctenostomes are the most uniserial of the groups, followed by anascaus, whereas cribrimorphs and ascophorans are overwhelmingly multiserial. Cribrimorphs are exclusively encrusting and ascophorans very largely so, whereas anascans and cyclostomes include as many or more erect as encrusting species. Excluding the uncommon cribrimorphs, anascans and cyclostomes are the most similar to one another in colony form, ctenostomes and ascophorans the most dissimilar.

Growth Form Model

What is the ecological and evolutionary significance of this variation? One approach to this question is to deduce the adaptive significance of different growth forms as strategies to reduce risk of mortality due to known ecological processes (Jackson 1985). Such a model artificially simplifies the problem to manageable proportions and leads to a series of testable hypotheses regarding the importance of bryozoan colony shapes to variations in their zooidal morphology, life history, and distribution.

The model used here considers risks of mortality for different growth forms living on hard substrata in terms of differences in their modes and directions of growth (Jackson 1979a, Coates & Jackson 1985).

Uniserial colonies are highly directional forms whose growth increases their chance of locating particular types of favourable habitats (spatial refuges) on an heterogeneous terrain. However, zooids in a chain lack neighbours on all sides for common protection against physical damage or enemies. Thus, the probability that any zooid will die in a given interval is high, but the probability that some small group of zooids somewhere in the colony will find a refuge and survive is also high. In contrast; multiserial growth suggests an increased commitment to survival in the immediate vicinity of larval attachment. Thus, multiserial colonies employ what might be termed a confrontational strategy in which survival is more dependent on maintenance of the genetic individual as a single integrated colony, or as a few integrated colonies, than in uniserial clones that may be highly fragmented. In this case the probability that any zooid will die is typically much less than in uniserial forms, but the chances of locating refuges by growth are lower as well.

Whether uniserial or multiserial, encrusting growth sets no special mechanical constrains for support of the colony. Thus growth is potentially infinite, although in practice limited by the boundaries of substrata, interactions with other organisms, and the vagaries of the physical environment. The costs of such structural freedom are greater vulnerability to all the potentially harmful processes that occur more intensively on the substratum—including sedimentation, overgrowth, and some forms of predation. In contrast, erect growth places stringent mechanical demands on skeletal systems for support and attachment. The potential benefits include greater access to resources above the substratum, escape from harmful processes restricted to the bottom, and enormous increase in surface area and biomass, and thus feeding and reproduction, for a given area of habitat.

Zooidal Characteristics and Colony Form

Bryozoan zooids are the basic units of construction from which colonies are made. Zooids vary considerably in size, shape, spacing, design, function, and integration, all of which should be related in some way with colony form.

Predicted Zooidal Characters

Survival of a genetic individual of a uniserial species often depends on survival of one or a few isolated zooids, In this case, chance of further survival should increase with zooidal size, implying selection for size increase. Zooids should be elongate or widely spaced to maximize location of spatial refuges. Integration of zooids and incidence of zooidal polymorphism should be low to increase chances that any surviving zooid can feed or reproduce. Moreover, when polymorphism occurs, different types of zooids should be distributed uniformly throughout the colony to increase the chances that all functions might survive.

Survival of a multiserial colony rarely depends on a single zooid. Potential size of a colony is largely independent of zooidal size so there should be no selection for zooidal size increase, at least so far as risk of mortality is concerned. In contrast, integration of zooids and incidence of zooidal polymorphism should be high, with clearly developed separation of functions (different kinds of polymorphs) within different regions of colonies, to increase chances of survival for the colony as a whole.

For both uniserial and multiserial growth, biomechanical influences on zooidal characters should differ greatly between encrusting and erect colonies. Because there are no special structural problems associated with encrusting growth, zooidal characteristics, and their expression within a colony, should be free to vary widely in response to biological factors affecting them. In contrast, mechanical constraints on erect growth, particularly that of arborescent species, sets strict limits on the distribution of types of zooids and special supporting structures independently of their potential advantages or disadvantages in biological interactions.

Intermediate growth forms between uniserial and multiserial, or encrusting and erect, are common, as in the fan-or ribbon-like colonies of Electra or Steginoporella that are intermediate between uniserial "runners" and multiserial "sheets". Zooidal characteristics of such colonies should be intermediate.

Observed Zooidal Characteristics

Zooidal morphology, spatial arrangement, and polymorphism are generally as predicted by the risk model in both cheilostomes and cyclostomes (Coates & Jackson 1985).

Results for cheilostomes are summarized in Table 7.1. Elongation and spacing vary as predicted by the model, with maximal values for uniserial species. However, zooidal size is least for these species because their zooids are extremely narrow. As the model predicts, the highest incidence of polymorphism is for multiserial species, most of which have both ovicells and avicularia, and lowest for uniserial species, most of which have only one.

There are also data on the relation of colony form to integration of zooidal feeding behaviour (Winston 1978, 1979). As predicted, integration is greatest for multiserial colonies, and least for uniserial ones, but there is no significant difference between encrusting and erect colonies.

Four cheilostome faunas from the Paleocene and Eocene (Canu & Bassler 1933, Cheetham 1963, 1966, 1971) were examined to determine to what extent zooidal characteristics predicted by the risk model were established at the beginning of the Tertiary radiation of the group (Coates and Jackson 1985). Results are very different from the Recent (Table 7.1): only zooidal elongation varies significantly between growth forms of Paleocene and Eocene cheilostomes, and the other three parameters do not show even a suggestion of the modern pattern. This increase in zooidal differentiation between growth forms suggests considerable, sustained selection for more than 50 million years. Similarly, the large overall increase in incidence of polymorphism (compare last two columns of Table 7.1) suggests a comparably important increase in integration of zooids, as had been previously indicated from somewhat different evidence for the entire span of cheilostome evolution.

Perhaps the best example of interdependence of growth form and zooidal characters is that of rigidly erect cheilostomes which first appeared in several independent lineages during the late Cretaceous (Cheetham 1971, 1986a). Binding, thickening, and joining together of zooids all depend on the existence of specific zooidal morphotypes, without which growth of erect, rigid cheilostomes could not occur. An obvious example is the

TABLE 7.1

Variations in Zooidal Size, Elongation, Spacing, and Incidence of Polymorphic Zooids in Relation to Colony Form among Recent and Early Tertiary (Paleocene, Eocene) Cheilostomes. Recent Data from Harmer 1926, 1957; Ryland and Hayward 1977; and Hayward and Ryland 1979; Fossil Data from Canu and Bassler 1933; Cheetham 1963, 1966, 1971. Size is the Geometric Mean of Zooidal Length times Width. L.W.: Elongation is LW; and Spacing is the Mean Distance Between all Orifices Surrounding a Zooid. Polymorphism was Scored 0=No Ovicells or Avicularia, 1=Avicularia or Ovicells Present, 2=Both Kinds of Polymorphs Present. F-Ratios and Probabilities for 1-Way Analysis of Variance of Parameter Values by Colony Form

Colony Form	*Size*		*Elongation*		*Spacing*		*Polymorphism*	
	Recent	*Fossil*	*Recent*	*Fossil*	*Recent*	*Fossil*	*Recent*	*Fossil*
Uniserial unilaminate	0.38	0.52	2.69	2.80	0.57	0.59	1.50	0.50
multiserial unilaminate	0.48	0.45	1.49	1.53	0.44	0.41	1.47	1.31
multiserial multilaminate	0.45	0.56	1.33	1.48	0.41	0.43	1.93	1.00

multiserial erect	0.43	0.44	1.86	1.90	0.38	0.43	1.79	1.16
uniserial erect	0.43	0.40	3.01	2.12	0.52	0.43	1.40	1.29
mean	0.45	0.45	2.00	1.76	0.45	0.43	1.54	1.22
number of species	462	138	462	138	414	127	522	143
F-Ratio	3.8	1.5	90	8.3	7.4	1.5	8.9	1.5
probability	.004	.21	.000	.000	.000	.21	.000	.21

benefit of frontal calcification of zooids associated with the extrazooidal thickening of branches towards the bases of arborescent colonies. More subtle is the relation between thickness of zooids at the growing tips of rigidly erect colonies and their inferred modulus of rupture. *i.e.*, resistance to bending as shown for 24 Recent and fossil species Modulus of rupture decreases significantly with increasing zooidal thickness ($r_s = -0.52$, $P < 0.01$).

Results for 33 species of Mediterranean tubuliporine cyclostomes. Uniserial species have larger and more widely spaced zooids, as predicted by the growth form model.

Environmental Distributions

Regardless of growth form, any sessile animal can survive where there is adequate substratum and food, the weather is favourable, competitors are absent, and potentially harmful processes like storms, sedimentation, or predation (often lumped together by ecologists as "disturbance") do not occur. With competition, or extreme or unpredictable variation in disturbance, however, different growth forms may be varyingly excluded or diminished in importance.

Predicted Distributions

According to the risk model, uniserial colonies should be favoured over multiserial ones at high levels of disturbance because they are more likely to have encountered a spatial refuge from disturbance. Moreover, growth requirements are less substantial than for multiserial forms, so that uniserial colonies are more likely to be able to grow large enough to reproduce before a calamity strikes. Also at high levels of disturbance, encrusting colonies should be favoured over erect ones because they can reproduce sooner (less growth required), and because encrusting colonies are harder to break or eat.

In contrast, at low levels of disturbance, multiserial colonies should be favoured over uniserial colonies because they are better competitors for space. They can also completely cover a substratum and thereby prevent invasion by foreign larvae. Similarly, erect colonies should be favoured over encrusting ones because they can get closer to food suspended in the water and, in so doing, reduce access to food by encrusting colonies beneath them, just as erect table corals inhibit those beneath them by shading (Stimson 1985).

The predictions of the model are easy to test in principle, but not in practice, because the distribution of intensity of competition and different forms of disturbance on the sea floor are not well known and are not independent. Important forms of physical disturbance include waves and currents, sedimentation, and extreme fluctuations in salinity or temperature, all of which decrease in intensity with increasing depth (Sverdrup *et al.* 1942). Their impact, however, may vary depending on a host of other factors. For example, deleterious effects of strong water movements should vary with the size or stability of the substratum, if it is small, slight movements may turn it over, whereas if the substratum is large or firmly attached this will not happen, and the water movements may even be beneficial as a source of extra food. Similarly, flexible or short-lived substrata, like fronds of algae and seagrass blades, are more easily moved about or destroyed than attached scleractinian corals. Indeed, the very abundance of suitable substrata, and thus that of bryozoans themselves, may depend upon the local physical hydrographic regime (Eggleston 1972a. Holme and Wilson 1985). Lastly, extreme variations in factors like salinity and temperature are much more likely in confined coastal areas than along open coasts.

Important forms of biological disturbance are predation and bioturbation. Like physical disturbance, both of these processes decrease with depth. The abundance of fishes, sea urchins, limpets, and other important grazers of hard substrata drops off considerably within 100—200 m of the surface (Lang 1974, Jackson and Winston 1982). Similarly, deposit-feeding animals that turn over sediments, and thus may disturb substrata suitable for bryozoans, like shell fragments or shells, are less abundant and turn over less sediment below a few hundred metres (Appendix 1 in Thayer 1983). Regardless of depth, the effects of both these processes may vary locally with the relative accessibility or susceptibility of habitats to these organisms.

In summary, it seems very likely that the total amounts of disturbance that may affect bryozoans fall off greatly with increasing depth, with increasing size or stability of the substratum, and away from estuaries and enclosed bays. Thus, according to the risk model, the ratios of uniserial to multiserial and encrusting to erect species should decrease in the same directions.

Observed Distributions with Depth

In our Atlantic sample, numbers of bryozoan species and all attached growth forms fall off precipitously with depth, except for the minute rooted forms. The ratio of encrusting to erect species falls off as predicted, with erect species becoming more diverse only below about 400 m. However, the uniserial to multiserial ratio shows little change in shallow depths, and then increases sharply below 1,000 m. which is contrary to prediction. Examination of the relative frequencies of the different growth forms with depth shows that this contrary result is due to an overwhelming predominance of uniserial erect species at great depths where they comprise half the fauna. In contrast, multiserial erect species show a more complex distribution, increasing in relative diversity with depth to about 1,000 m, and then falling off again to about the same percentage as at the surface.

Some insight into these patterns emerges from information on the attachment and construction of erect species which may be:

(a) cemented or rooted,

(b) inflexible or flexible (independent of jointing), and

(c) unjoined or jointed.

Between the surface and 150 m, there is a sharp, more-or-lass parallel increase in the proportions of cemented, inflexible, and unjointed species. This is followed by a gradual decline to 1,000 m, and then an even steeper decline below. Moreover, this pattern largely reflects that of multiserial erect species, while the pattern for uniserial species is much less pronounced. Cemented attachment, inflexible construction, and lack of joints combine to make a rigid structure compared with colonies that are rooted, flexible, or jointed. This rigidly erect construction is less frequent in shallow waters, presumably due to problems of breakage by water movements (Holme and Wilson 1985) and predators (Vance 1979). On the other hand. its subsequent precipitous decrease with depth (beginning for cemented species between 300 and 400 m) must be due to other causes, most probably decreased availability of suitably stable substrata, and perhaps frequent shifting of sediments from turbidity flows. The impression is that the deep sea is not so environmentally stable as is often suggested, at least with regard to substrata for sessile attachment.

There are too few quantitative abundance data for living colonies to compare with these patterns. Trawl and camera surveys in the North Sea (Dyer *et al.*, 1983) and English Channel (Holme and Wilson 1985) have revealed abundant populations of foliaceous erect species, such as Flustra foliacea and Pentapora foliacea, but other smaller or less conspicuous bryozoans were not examined. The latter study is particularly valuable for clearly showing the complex interacting effects of water movements, in this case tidal currents up to 160 or even 250 cm sec^{-1}, and the effects of transported sediments. Stable, hard bottoms exposed to such water movements, but not subject to scour or burial by sediments, are covered by abundant encrusting sponges, bryozoans, ascidians, etc., including large numbers of rigidly erect Pentapora foliacea and Omalosecosa ramulosa. In contrast, similar bottoms subjectto scour and potential for burial lack sponges and rigidly erect bryozoans, but support numerous multiserial erect, flexible colonies of cemented Flustra foliacea and rooted Vesicularia spinosa. Foliaceous species also dominate several faunas below

30 to 100 m off New Zealand (Probert and Batham 1979. Bradstock and Gordon 1983) and in the Gulf of Mexico (Canu and Bassler 1928. J.E. Winston, personal communication).

Rigidly erect cheilostomes are abundant below 50m in the Ross Sea and several coastal bays along the Antarctic coast, where they may comprise one-third or more of all the bryozoan species present, and on the Grand Banks off Newfoundland (Powell 1968). However, they do not appear to be at all common in tropical seas shallower than 100m, particularly in reef environments (Jackson 1984, unpublished). In contrast, small colonies of rigidly erect species are the predominant growth form among cyclostomes in the majority of habitats in depths less than 100 m in the Mediteranean (Harmelin 1975, 1976). Large, rigidly erect cyclostomes do occur today, but much less frequently than cheilostomes.

Distributions on Different Substrata

Bryozoans grow on an enormous variety of substrata. To simplifv analysis, we grouped these into three categories:

(a) live, mostly soft or flexible—including all algae (except crustose corallines), seagrasses, hydroids, ascidians, etc. —and carapaces or shells of live crustaceans, molluscs, or hermit crabs;

(b) mostly dead, hard substrata—including shell fragments, shells, pebbles, stoges, concretions—and living or dead crustose corallines; and

(c) dead surfaces of scleractinian corals.

The reason for separating corals is that they are typically much larger and longer lived than the other hard substrata, and are commonly cemented to the bottom. Stability of the three groups of substrata is inferred to increase in the order presented. Rock walls are even more stable than corals, but could not be included for lack of data.

The shaded bars are for species in our entire Atlantic sample, the open bars for cheilostomes from south-east Florida and the Caribbean in depths less than 100 m, for which ecological data are more complete (Winston 1982, 1984a, Jackson 1984). Stability increases from left to right. The pattern for uniserial versus multiserial species is as predicted. Uniserial species comprise almost half the fauna on algae and seagrasses at many sites, and are common on shell fragments and small bits of rubble, but are rare on corals. In contrast, the ratio of encrusting to erect species is opposite that predicted by the risk model. This is due to the great predominance of uniserial erect species on algae and seagrasses, and the rarity of any erect species on corals (note especially the Caribbean data), Predation upon sessile organisms is intense on coral reefs (Sammarco 1980, Jackson and Winston 1982, Jackson and Kaufmann 1986), and multiserial erect species are virtually absent except for a few cryptic habitats. Thus predation appears to be more important than substratum stability for the occurrence of erect species, at least in shallow water (see also Vance 1979).

Distributions with Salinity

Table 7.2 contrasts the distributions of taxa in our Atlantic survey with Winston's (1977a) compilation of species reported from waters of diminished salinity world-wide. As predicted by the risk model, the two overwhelmingly multiserial groups, ascophorans and cyclostomes, are the most diminished in brackish waters, while the predominantly uniserial ctenostomes are diminished least. In contrast, there is no relation between the overall ratio of encrusting to erect species in each taxon and their representation in brackish water.

We further examined the relation between growth form and salinity using Winston's (1982) survey of the fauna from south east Florida. There are 21 species from waters of diminished salinity, and 63 from normal marine waters. As predicted, the ratio of uniserial to multiserial species is extremely high in brackish water—2.4, whereas it is only 0.28 elsewhere.

Contrary to prediction, however, the ratio of encrusting to erect species is lower in brackish water (1.2) than elsewhere (5.6). This higher proportion of erect species in brackish water is of special interest because of the decrease in abundance and diversity of predators in estuaries (Jackson 1985), a pattern which further supports the idea that predators play a major role in determining the relative abundance of erect growth forms in shallow water.

TABLE 7.2

Taxonomic Distributions of Species Reported from Brackish Water World-wide (Winston 1977a) and for all Species from all Environments in our Compilation of Atlantic and Western Mediterranean Bryozoans. Data are Numbers of Species and Percentages in Differences in Parentheses. Taxonomic Proportions are Highly Significant ($\chi^2=108$, P. <0.001)

Taxon	*Brackish water*	*All Atlantic environments*
Anasca*	92 (45%)	463 (32%)
Ascophora	43 (21%)	683 (48%)
Ctenostomata	52 (26%)	99 (7%)
Cyclostomata	16 (8%)	185 (13%)
total	203 (100%)	1430 (100%)

*Includes cribrimorphs.

Growth Forms in the Fossil Record

Growth forms of many fossil bryozoans are different in important details from Recent species, but they can still be analyzed in the same general way.

Uniserial Versus Multiserial Fossils

Uniserial encrusting bryozoans have been common but not diverse on shells and similar substrata throughout the history of the phylum. Examples include the widespread occurrence of uniserial cyclostomes such as Corynotrypa and Stomatopora on shell debris from the Ordovician to the end of the Mesozoic and the earliest cheilostome, Pyriporopis. In contrast, uniserial erect cheilostomes are uncommon as fossils, probably due to problems of preservation of their weakly calcified skeletons. Univerial erect forms comprise 16 per cent of the 1,430 Recent species in our survey, but only 7 per cent of the Paleocene and Eocene species in Table 7.1. Similarly, uniserial encrusting species are common in Maastrichitian hard-grounds and seagrass deposits, but uniserial erect species are not (Voigt 1981). Consequently, compilation of ratios of uniserial to multiserial fossil species seems fruitless and potentially misleading.

Encrusting Versus Erect Fossils

There was a complete reversal in the diversity of encrusting versus erect bryozoan species between the Paleozoic and post-Paleozoic. Overall, the median ratio of encrusting to erect species for the Paleozoic is 0.10, and for the post-Paleozoic is 0.97 (Mann—Whitney U Test, $P=0.0009$). Paleozic faunas are overwhelmingly dominated taxonomically by erect species, and appear to have become increasingly so after the Silurian (Mann's test for trend, $T=3.20$, $P=0.007$). In contrast, post-Paleozoic faunas include, on average, equal numbers of encrusting and erect species and, since the Jurassic, have become dominated by encrusting species, especially in the Neogene.

The pedominance of encrusting species in Cenozoic seas is in stark contrast to the impression gained from the best-known fossil bryozoan faunas from the Ordovician onwards. Rigidly erect cheilostomes volumetrically dominate the majority of bryozoan-rich deposits of Upper Cretaceous to Recent age personal communication, Lagaaij and Gautier 1965, Labracherie 1973, Thomsen 1976, Balson 1981, Cheetham and Thomsen 1981, Cuffey *et. al.* 1981). Likewise, rigidly erect stenolaemates usually dominate Paleozic and Mesozaic deposits.

The taxonomic and volumetric dominance of rigidly erect multiserial cheilostomes and cyclostomes on Cretaceous seagrasses is particularly interesting. More than 70 of 130 species found on the Masstrichtian seagrass Thalassocharis formed rigidly erect colonies. In contrast, only 6 of 110 species living today on Posidonia in the Mediterranean have this morphology (Harmelin 1976); the rest are encrusting or grow as typically uniserial, flexibly erect colonies.

For a variety of reasons, these differences between volumes of rigidly erect colonies in fossil and Recent faunas could be more apparent than real. Ecologists studying living bryozoans have worked primarily in the intertidal zone or at depths less than 20m. Deeper water faunas have been investigated primarily by systematists provided with little ecological information, making Winston's (1983) study of inferred patterns of growth, reproduction, and mortality of Antarctic rigidly erect species unusual. Thus, Recent faunas dominated by these forms may be more common than now realized.

On the other hand, many rigidly erect bryozoans probably grow fast and suffer extensive breakage and partial mortality as do the majority of branching corals (Highsmith 1982). If so, great accumulations of fossil sketetal material may suggest abundances of erect bryozoans far greater than they were in life, much like the problem of discarded moults of trilobites. Indeed, much of the sea bottom visible in photographs for Antarctica seems rather barren of erect bryozoans, even though the phatographs were specifically chosen to illustrate these organisms; yet the sediment on which the bryozoans are growing is rich in bryozoan debris (Winston 1983).

There is other evidence, however, that suggests more fundamental differences between ancient and Recent distributions of rigidly erect species. Modern depth distributions of growth forms in shallow water, particularly the upper limit of rigidly erect species, are very different from Paleozic distributions. Then, these species were the dominant forms on hard substrata, both in diversity and estimated abundance, throughout epicontinental seas. McKinney 1979, McKinney and Gault 1980, Karklins 1984, McKinney *et al.*, 1986b). Many were quite large, commonly reaching 10-20 cm in height or diameter, and occasionally 1—3 (*e.g.*, Ross 1972, Smith 1981). Similarly, large erect forms dominated shallow level bottoms. Bryozoans such as Archimedes, Fistulipora, Tabulipora, and various Ordovicians trepostomes built up enormous populations on lagoonal floors and shoals (Cuffey 1967, Ross 1970 McKinney 1979, 1983, Ettensohn *et al.* 1986). These environments today are dominated by branching corals, or may lack erect animals entirely.

Physical processes in shallow waters are unlikely to have changed radically during the Mesozoic, certainly not enough to push multiserial erect species 100 m deeper, which strongly suggests a biological explanation. The most likely candidate is the dramatic escalation in effectiveness of the grazing, browsing, and gouging machinery of potential predators upon bryozoans throughout the Mesozoic (Vermeij 1977, Steneck 1983). Particularly pertinent are the appearances of sea urchins with well-developed Aristotle's lanterns in the Jurassic and Cretaceous, and of fishes capable of excavating calcareous material in the Eocene. A good example is the sea urchin. Centrostephanus coronatus, that feeds preferentially on erect bryozoans and other erect sessile organisms on shallow rock walls off southern California. Consequently, erect animals are rare except when the urchin has been removed experimentally or is naturally absent or rare (Vance 1979). Abundances of all these predators drop off sharply with increasing depth.

The present is not always the key to the past. The presence of rigidly erect colonies, particularly those with radial branches, commonly had been accepted as suggestive of relatively "deep" offshore shelf deposits in studies of Palcozoic bryozoans (*e.g.*, McKinney 1972) until numerous instances providing irrefutable evidence of very shallow, often almost intertidal, occurrences of these forms were recognized. Apparently, shallow shelf environments are more disturbed by predators than in the Paleozoic. Thus, the deeper modern distribution of rigidly erect species conforms in a general way with the predictions of the risk model.

Growth forms as Adaptive Strategies

So far we have considered different growth forms as alternative ways of making a living, without explicitly considering species that grow in more than one shape. However, 340, or 24 per cent, of the species encountered in our survey have this potential. Such variability most commonly involves transitions between uniserial encrusting species and either multiserial encrusting species ("runner" to "sheet") or uniserial erect forms ("runner" to "vine"). Also common are transitions between multiserial entrusting and foliaceous rigidly erect growth, which merely involves raising of the growing edge away from the substratum.

Variations in encrusting growth patterns within a species seem readily explicable in terms of plastic responses of changing environmental conditions. The best example is that of Conopeum tenuissimum, whose growth changes from predominantly uniserial to multiserial with increasing quantity and quality of food. Another example, perhaps also related to food availability, is the ribbon-like growth of Steginoporella sp. under deep corals and its broad fan-like growth under shadow corals. Similar variations occur in colony morphologies of rigidly erect species (Harmelin 1973, 1975).

Cheilostomes capable of frontal budding become multilaminar once this process begins, but the timing is so variable that some colonies remain unilaminate throughout their lives, whereas others begin multilaminate growth almost immediately. Early frontal budding may be a response to restrictions to continued horizontal growth, such as the limits of a substratum, or to interference by other species, as in Antropora tincta (Buss 1981b). In other cases, such as Gemelliporidra belikana that lives in heavily grazed habitats on coral reefs, frontal budding always begins before the colony has reached ten zooids (M. Gleason and J. Jackson, unpublished data). Apparently, the increased relief of these heavily calcified colonies helps to resist grazing, for they are little affected.

The formation of erect foliaceous colonies from encrusting bases may be controlled by predation orother forms of disturbance. A good example of this is the sponge Cliona celata off North Carolina (Guida 1976). Under conditions of heavy grazing by sea urchins, this excavating sponge lives only in borings within calcareous substrata. In the absence of predators, however, it grows over the outer surface of its substratum and eventually forms erect branches. The same is probably true of western Atlantic Steginoporella spp. and Stylopoma spongites that usually grow only as encrusting colonies on coral reefs, but as foliaceous erect colonies offshore where there are fewer durophagous predators (J. Jackson, unpublished data).

All of the above changes in form can be explained as plastic developmental responses to changes in extrinsic environmental factors, including physical processes, food, competitors, and predators. These are the "unstable" species of Stach (1937). In contrast, most rigid, arborescent species are stable in that they begin life with encrusting growth, but very soon develop a more-or-less fixed vertical pattern, disrupted only by injury to the colony. The selective basis for this inflexible extra investment in support presumably involves benefits to the colony that accrue only in later stages of development (Jackson 1979a). Clearly, specialization in growth form decreases ecological options, thereby fixing the life history strategy of the species in many other ways. These ideas are pursued in the next chapter.

8

Ecology of Insects

Origin and Evolution of Insects

Evolutionary change may be viewed as occurring in three successively longer time frames: microevolution, the changes that occur in populations usually in response to changes in selection pressures, such as industrial melanism and the development of insecticide resistance in a population; speciation, the changes that typically occur over much longer time spans than microevolution and that result in reproductively isolated, new species; and finally macroevolution, the major phylogenetic patterns that develop over wide spans of geological time.

The objective of this section is to outline the major macroevolutionary events that are thought to have occurred among insects and their relatives. The discussion includes consideration of the origin of the arthropods, origin of the insects, and phylogeny within the class Insecta.

At least a rudimentary knowledge of evolutionary theory is assumed in this discussion. Readers unfamiliar with natural selection, speciation, the concept of geological time, and so on, should consult a good general biology or zoology text or any of several excellent texts available which deal specifically with organic evolution (see Selected References).

Comparison of present-day (extant) forms has led to the recognition of structural and developmental affinities among the Annelida, Arthropoda, and Onychophora.

Annelida. Annelida, the segmented worms, apparently evolved from the ancient protostomes, which also gave rise to the mollusks (Meglitsch, 1967). Annelids are soft-bodied, elongate, cylindrical, bilaterally symmetrical animals. Their bodies are composed of repeating segments or metameres capped anteriorly by the prostomium or acron and posteriorly by the periproct or telson. The mouth is on the venter between the prostomium and first metamere. The anus opens from the periproct. Some annelids have several bilateral pairs of segmental appendages called parapodia, which arise as evaginations of the body wall. Segmentation of the body is evident in nearly all tissues save the alimentary canal, which is essentially a longitudinal tube running from the mouth to the anus. The circulatory system is a closed system of tubes, usually one dorsal longitudinal and one ventral longitudinal vessel, connected by lateral trunks that provide vascularization for the various body tissues. The blood commonly carries a respiratory pigment. The body cavity is a coelom divided into segmental compartments. Excertion is accomplished by paired segmental nephridia, tubes that communicate between the coelom and the exterior. Ventilation is cuticular or by means of gills evaginated from the body wall. The nervous system is composed of an anterior prostomial ganglion which lies dorsal to the alimentary canal. It is connected by a pair of bilateral connectives to a ventral chain of paired segmental ganglia linked by paired longitudinal connectives. Some annelids have a trochophore larval stage, which has played an important role in assessments of the evolutionary relationships of annelids with members of other invertebrate phyla.

Arthropoda. Arthropoda, the joint-footed animals, are a highly successful diverse group of animals unrivaled by other animals in diversity of structure and function. They are found in nearly every imaginable ecological niche. A feature that has played a central role in their success is their relatively impermeable exoskeleton. Such an impermeable covering was a prerequisite for invasion of the terrestrial environment, which poses a continuous challenge to the waterholding capabilities of all animals and of arthropods in particular because of their generally small body size. Additional adaptations are discussed later in this chapter.

Arthropods have segmented bodies composed of chitinous exoskeletal plates of varying degrees of hardness, which are separated by less-hardened "membranous" areas. Paired segmental appendages are usually present on at least some body segments. Arthropods share several fundamental traits with annelids. Particularly outstanding are their segmented bodies and the ventral chain of segmental ganglia. On the other hand, the arthropods have an open circulatory system, and the body cavity is a hemocoel; coelomic sacs appear, during embryogenesis, but do not form the body cavity. Also most arthropods ventilate by means of a tracheal system and have well-developed appendicular mouthparts. For more information on the arthropods, see Boudreaux (1979), Clarke (1973), and Manton (1977).

Extant arthropods include the chelicerates, crustaceans, myriapods, and insects.

The chelicerates include the spiders, mites and ticks, scorpions, solpugids, and their relatives (class Arachnida). Horseshoe crabs (Xiphosura, *e.g.*, Limulus), sea spiders (Pycnogonida), water bears (Tardigrada), and tongueworms (Linguatulida or Pentastomida) are variously classified, some authors including them with the chelicerates. Chelicerates are characterized by the presence of six pairs of appendages; the anterior pair are called chelicerae and are jawlike in more primitive forms. They have been modified in a variety of ways in more advanced forms. The second pair of appendages are called pedipalps and also vary in structure. The remaining four pairs of legs generally have a locomotor function. Antennae are lacking, and there are usually two body regions; a fused head and thorax (cephalothorax) and an abdomen. The appendages are borne by the cephalothorax.

The crustaceans (cryfish, lobsters, crabs, shrimp, barnacles, copepods, sowbugs, and so on) have two pairs of antennae and a pair of mandibles. The crustaceans commonly have two body regions: a cephalothorax, which is covered by a carapace, and an abdomen. These arthropods are mostly freshwater or marine animals.

Myriapods are cylindrical, have a well-defined head bearing primitive mandibulate mouthparts, and have a trunk with nine or more bilateral pairs of legs. There are four myriapod classes: Diplopoda (millipeds), Chilopoda (centipedes), Pauropoda (pauropods), and Symphyla (symphylans). All are terrestrial animals.

The insects possess a head, thorax (usually with wings and legs), and an abdomen. This is the only group of invertebrates that contains members capable of active flight. Insects are primarily terrestrial with many secondarily aquatic forms.

A major group of extinct arthropods, the trilobites, have played an important role in the study of arthropod evolution. These animals became extinct or very rare in the Permian period and therefore are known only from the fossil record. They were marine animals, and their bodies were made up of three tagmata: head, thorax, and pygidium. The name Trilobita refers to these three tagmata. The head and first few thoracic segments were covered by a carapace, most of the body segments had a bilateral pair of jointed legs, and the head carried a single pair of antennae.

Onychophora. Onychophorans are a very small group of terrestrial animals, but are significant in that they show morphological affinities with both annelids and arthropods. These animals have elongate, wormlike bodies with several bilateral pairs of lobelike legs. They are not segmented as adults, but are clearly segmented during their embryogenesis. They have paired nephridia (excretory organs) along the body that open near the legs bases. This fact, together with other morphological characters, suggests an evolutionary relationship with the annelids. On the other hand, onychophorans share several characters with arthropods, including a chitoproteinous cuticle (Hackman and Goldberg, 1975), an open circulatory system with a hemocoel, tracheal ventilation, and appendages associated with the mouth. Onychophorans have been variously classified as annelids, arthropods, or as a separate phylum.

Origin of Arthropoda

The basic similarities among annelids, arthropods and onychophorans strongly suggest common ancestry at some point in metazoan evolution. Some authors have, in fact, classified all three groups as members of superphylum, Annulata (Snodgrass, 1952). Sharov (1966), in a similar vein, supports Cuvier's* concept of the phylum Articulata with annelids, onychophorans, and arthropods being recognized as separate subphyla. The details of the evolutionary relationships among these three groups are purely hypothetical. Unfortunately, knowledge of the fossil record is of little help, since each group is distinctly defined in the Cambrian strata, from which the oldest 'good'' fossils come. This, of course, means that the supposed common ancestor arose in pre-Cambrian times. There is, however, general agreement that the common ancestor was an annelid or annelidlike creature.

Some have suggested that the primitive arthropod evolved directly from a primitive polychaete annelid (Meglitsch, 1967). Snodgrass (1952), later supported by Sharov (1966), conceived of a segmented wormlike animal that produced two branches, one ultimately giving rise to the polychaete annelids and the other to a form with paired, undifferentiated lobelike legs

*Georges Cuvier (1769-1832) was among the founders of paleontology, the study of the fossil record.

(lobopod). He envisioned the lobopods as subsequently branching into arthropodan and onychophoran lines.

Once established, members of the phylum Arthropoda are viewed by some as having branched out into three groups (subphyla): the trilobites (Trilobita) the chelicerates (Chelicerata) and the mandibulates (Mandibulata), which include the Crustacea, myriapods, and insects.

There is disagreement as to whether the arthropods actually constitute a single phylum (are monophyletic), sharing a common arthropodan ancestor, or are actually two or three groups (are polyphyletic) of distantly related animals in which major arthropodan traits, especially jointed appendages but also many other structures, evolved convergently (Scudder, 1973; Boudreaux, 1979). Recent cases for the latter point of view have been made by Tiegs and Manton (1958), Manton (1964, 1970, 1972, 1979), Cisne (1974), and Anderson (1973, 1979). On the basis of mandibular structure and movement and locomotor limbs, Manton recognizes three groups that became "arthropods" convergently the trilobites and chelicerates; (2) the Crustacea; and (3) the Onychophora, myriapods, and hexapods (insects). Anderson (1973, 1979) corroborates this point of view embryologically. The trilobites, chelicerates, and crustaceans have primitively biramous appendages and gnathobasic mandibles. The form of the gnathobasic mandibles and biramous appendages in the trilobites and chelicerates is fundamentally different from those of the crustaceans. On these and other bases, the trilobites and chelicerates are viewed as evolving independently. The Uniramia all have similar uniramous limbs and bite with the tips of the mandibles: thus they are considered to constitute a monophyletic group. Cisne (1974) studied parts of the internal anatomy of pyritized trilobites using X rays and provides evidence in support of a polyphyletic origin of arthropods. However, he views the trilobites as a link between the chelicerates and crustaceans and hence recognizes two arthroped lines.

Although there is substantial evidence for a polyphyletic origin of arthropods, acceptance of this point of view requires acceptance of the independent, convergent origin of several characters; a hemocoel: paired, jointed locomotor appendages; tracheae; Malpighian tubules; compound eyes; a chitinous cuticle, and so on. Some experts find this very difficult to accept. In a recent multiauthored volume entitled Arthropod Phylogeny (Gupta, 1979), a monophyletic origin of arthropods is supported by studies of the embryology of the head, olfactory mechanisms and sensilla, eye structure, visceral anatomy, intersegmental tendon systems, and sperm ultrastructure. Boudreaux (1979) also favours a monophyletic explanation of arthroped origin and provides an especially useful discussion of arthropod phylogeny. Unfortunately, the fossil record has not provided sufficient information to resolve the issue of a mono-versus polyphyletic origin of the arthropods.

ORIGIN OF INSECTS

The class Insecta is generally considered to have evolved from a myriapod or protomyriapod of some sort during the Devonian period. Based on differences in mandibles and mandibular movement, Manton (1964) concluded that insects are not direct descendants of myriapods and that these two groups are best looked upon as sharing a common ancestor.

A series of diagrams originated by Snodgrass (1935) and modified by Ross (1965) is helpful in a very simplified way in visualizing the hypothetical origin of the insects. The first represents the segmented, legless, wormlike annelid or annelid-like stage. The undifferentiated body is composed of a series of somites or metameres capped anteriorly by the prostomium (acron) and posteriorly by the terminal body segment, the periproct (telson). The mouth is lacated between the prostomium and the first body segment; the anus opens in the periproct. It epresents the evolutionary stage in which were developed paired, bilateral, lobelike appendages on the somites as well as a pair of simple eyes on the prostomium and antennae on the second somite. The appendages on the first somite have been lost altogether. This level of organization is onychophoranlike. The next stage represents a protomyriapod/

protoinsect in which arthropodization has occurred; that is, the bilateral appendages of each body segment have themselves become segmented. The appendages of somites 4, 5, and 6 have been reduced and have moved into close association with the primitive head, becoming involved with the manipulation of food. The appendages of the hindmost somite have become sense o gans and no longer function in locomotion. As mentioned earlier, there are varying opinions as to the origin of the myriapod and insect groups, but there is little doubt of a close relationship between these two groups.

The evolution of the insect head from the prostomium plus four somites, which is implied has been disputed. Rempel (1975) reviews the evidence regarding various theories of origin of the insect head and supports the one depicted, where the head originates from the fusion of the prostomium and six somites.

Insect Phylogeny

Carpenter (1953, 1977) recognizes four major stages in the evolution of insects. The first was the appearance of primitive wingless insects, which probably resembled contemporary Archeognatha and Thysanura. These primitive apterygotes are thought to have arisen during the Devonian period. The fossil record of apterygates is poor, probably because of their relatively soft bodies. Other insect groups with harder body parts and particularly with wings have been more amenable to fossilization processes. Despite lack of extensive fossil evidence, insects are generally thought of as originally having been wingless.

The second major step was the development of wings, which is thought to have occurred prior to the Lower Carboniferous period. Insects were the first animals to invade the air, doing so some 50 million years before the reptiles and birds. Since predatory amphibians, reptiles, and arthropods were abundant at this time, wings no doubt provided a strong selective advantage by facilitating escape. The early winged insects had very simple wing articulations that allowed flight, but at rest the wings were held out from the body, not being flexed out of the way over the abdomen. These paleopterous ("primitive-winged") insects became the dominant group of insects during the Carboniferous and Permian periods, but are represented today by only two orders, Odonata and Ephemeroptera.

A number of hypotheses have been advanced to explain the origin of wings in insects. These hypotheses are based mainly on studies of extent forms, the fossil record providing little definitive information since most of the earliest insect fossils known already had wings. It is generally agreed that wings and their control and propulsive mechanisms have been superimposed on segments originally adapted to walking. Wings have been added without the loss of legs. This is different from the vertebrates, in which walking appendages and the associated nerves and muscles have become the organs of flight.

A widely accepted theory of the origin of insect wings views these structures as having arisen as bilateral expansions (paranotal processes) of the thoracic nota (Muller, 1879; Hinton, 1963a, 1977; Flower, 1964; Hamilton, 1971). Hinton (1963) and Flower (1964) envision paranotal expansions serving first as surfaces that enabled insects to control body attitude when falling from vegetation, ensuring that they would land in an upright position and could immediately run to shelter if being pursued by a predator. Subsequently, ith further lateral expansion of the paranotal processes, gliding from high vegetation or even a running jump became possible. It is not difficult to imagine a subsequent gradual development of an articular and neuromuscular arrangement by which the tilt fore and aft of the paranotal processes could be varied, thereby exerting a degree of glide control. The next step would have been the development of basal articulations and neuromusculature that allowed the wings to exert both controlling and propulsive forces. It is possible that the original paranotal processes first functioned as other than organs of flight. For example, it has been suggested that they were originally involved in courtship displays by male insects (Alexander and Brown, 1963).

Recently, Wigglesworth (1976c) has renewed support for a theory, advanced in the late 1800s, which views insect wings as having orignated from tracheal gills in a ptimitive aquatic form. Gills in aquatic insects are commonly articulated, contain branching tracheae, and are movable. For example, mayfly larvae are able to wave their gills to and fro, a movement that brings oxygenated water closer to the body. Gills may also aid in swimming.

The third major evolutionary advance appears to have occurred during the Lower Carboniferous. This was the development of a wing-flevion mechanism, which enabled insects to flex the wings posteriorly over the abdomen, the neopterous ("new winged") condition. This added the advantage of being able to run and hide from predators and to move into niches inappropriate for forms with continuously outstretched wings. The neopterous insects subsequently radiated rapidly and became the dominant group of insects, as they are today. They comprise 90 per cent of the contemporary orders and 97 per cent of thetotal number of species (Carpenter, 1953). Only apterygotes and the Odonata and Ephemeroptera are not neopterous orders.

The final major evolutionary step was the development of complete metamorphosis, probably during the Upper Carboniferous. The evolution of complete metamorphosis enabled insects to benefit from the favourable aspects of different habitats. For example, an insect with aquatic larval stages and terrestrial-aerial adults would benefit from the abundant and readily available nutrients of the aquatic habitat and later from the potential for dispersal (and hence enhanced genetic mixing) offered by the terrestrial-aerial habitat.

Hinton (1977) discusses several insectan features, in addition to those recognized by Carpenter (1977), the origin of which he considers as major events in the evolution of insects. Among these are the evolution of a tracheal system, a relatively impermeable cuticle (mentioned earlier), and the fat body. These major adaptations are characteristic of most terrestrial arthropods and likely preceded the evolution of insects as such. They are therefore perhaps best viewed as preadaptations, which set the stage for the major evolutionary events just outlined.

The evolution of the tracheal system enabled insects (and other terrestrial arthropods) to localize respiratory and moisture exchanges with the environment as oppnsed to these exchanges occurring over the general body surface. This was followed by the development of a relatively impermeable cuticle (see "Permeability", "Arthropods" earlier in this chapter), a feature that enabled insects (and other arthropods) to radiate widely into terrestrial habitats. Evolution of spiracular closure mechanisms also deserves mention in this context (see "Ventilatory System"). These developments made possible the evolution of flight by counteracting the drying tendency of the ambient air. Hinton views the fat body, particularly with regard to the storage of fat, as a major terrestrial adaptation that allows insects and other arthropods to withstand comparatively long periods without water. He points out that the total amount of fat in arthropods is greater in terrestrial forms than in aquatic forms.

Hinton (1977) also views the size range of insects as a particular advantage, pointing out: "Insects are large enough so that they are not imprisoned by surface tension forces; many make great use of these during locomotion. At the same time they are small enough so that they can fall to any distance without injury." Janzen (1977) sees a great ecological advantage to the typical small size of insects: "I think that there are so many species of insects because the world contains a very large amount of harvestable productivity that is arranged in a sufficiently heterogeneous manner that it can be partitioned among a large number of populations of small organisms."

During the course of their evolution, insects have clearly influenced and been influenced by other organisms. See the section "Adaptations Associated with Interspecific Interactions" for examples of close (coevolutionary) ties between insects and other organisms. For extensive information on the evolutionary relationships between insects and plants, see Smart and Hughes (1972).

Based on Carpenter's four major evolutionary stages in insects, the class insecta may be divided into groups as follows:

Apterygotes—primitively wingless insects

Pterygotes—winged insects

Paleopterous exopterygotes—wing-flexion mechanism lacking; simple metamorphosis

Neopterous exopterygotes—wing-flexion mechanism present; simple metamorphosis

Neopterous endopterygotes—wing-flexion mechanism present; complete metamorphosis

Although it has not provided definitive answers, the geological record has been of value in piecing together many aspects of insect phylogeny and in evaluating the evolutionary associations between insects and other organisms. Since fossilization occurs only under certain very uncommon circumstances, the fossil record for most organisms is fragmentary. Insect fossils consist mainly of wings, but various other body parts are also commonly represented. In addition to representation of most extant orders (save Zoraptera, Grylloblattodea, Mallophaga, and Anoplura) in the fossil record, 52 extinct orders of insects have been described by paleoentomologists (see Carpenter, 1977; Smart and Hughes, 1973). Many of these orders are based on very limited data, such as isolated wings or wing fragments. Carpenter (1977) reduces the number of recognized extinct orders significantly by applying the following criteria: "My acceptance of an extinct order requires the knowledge of both fore- and hindwings (in the case of Pterygota) and the nature of the head, including mouthparts." On these bases 10 extinct orders can be recognized.

Apterygotes
- order Monura

Pterygotes
- Paleopterous exopterygotes
  - order Palaeodictyoptera
  - order Megasecoptera
  - order Diaphanopterodea
  - order Protodonata
- Neopterous exopterygotes
  - order Protorthoptera
  - order Caloneurodea
  - order Miomoptera
  - order Protelytroptera
- Neopterous endopterygotes
  - order Glosselytrodea

The spans of geological history known to have been occupied by members of these extinct orders and their possible relationships to main insectan lines.

9

Aquatic Insects of Soil and Water

Insects that spend part or all of their lives in soil or water exhibit special structural and behavioural adaptations to the physical, chemical, and biotic conditions found in each. In this chapter we shall examine some of the distinctive features of these media as habitats for insect life and discuss how these features are related to insect biology. We shall also note the important role of insects as decomposers in the renewal of these vital natural resources. The chapter concludes with a discussion of the early evolution of insects in soil and water.

Soil Insects

Soil is the medium that connects all habitats of the land, from the intertidal to the alpine. It is also a distinct realm and one on which green plants and ultimately all forms of terrestrial life depend. The soil surface and overlying litter of plant debris are inhabited by at least some members of nearly all insect orders, as well as by other terrestrial arthropods.

Beneath the surface the variety of species markedly declines, but millipedes, mites, springtails, beetles, termites, ants, and fly larvae are still represented, sometimes abundantly. Relatively few kinds of insects inhabit caves, but they are nonetheless interesting because of their unique adaptations.

Soil as a Habitat for Insects. The process of soil formation leads to horizontal layers that are readily visible in a cross section such as a road cut. On the surface is usually a duff, or litter of whole or partly decayed leaves and other plant parts. The nature of the filter varies with moisture and the covering vegetation, *i.e.*, grassland, coniferous forest, deciduous forest, or lichens and mosses. The soil's uppermost layer, or topsoil, is darkened by a complex, organic substance called humus. Topsoil is the place where most plant roots and soil organisms occur.

The transformation of the litter into humus is accomplished in stages. Arthropods contribute significantly to the process by ingesting organic matter at the surface, mechanically breaking it down, and defecating deeper in the soil. Their excavations also mix the surface layer between litter and soil. Bacteria and fungi in the moister regions below then act on the fragments, now with greater surface area, to produce humus. The deeper subsoil, if present, is paler, being only slightly enriched from above, and grades below into the unaltered rock of the earth's crust.

Cracks, old root channels, and tiny cavities between soil particles create pore spaces in soil. The total pore space of topsoil is usually about a third of the soil volume, but in some places may exceed half the volume. At greater depths, the soil is more compact and has smaller pore spaces. Depending on the nature of the soil particles, rainfall, groundwater, and drainage, the pore spaces are filled partly by air and partly by water held by capillary forces. Soil air is usually saturated with water vapour except near the soil surface. In well-drained topsoil, oxygen in soil air is replenished by gaseous diffusion from the atmosphere. Soil water derives its oxygen from soil air. Oxygen, however, diffuses 10,000 times slower in water. The consumption of oxygen in soil water by bacterial decomposition of organic matter can create anaerobic conditions in as short a distance as 1 mm from a water-air interface. Deeper in the soil gaseous diffusion from the atmosphere decreases. Here the respiration of plants and animals reduces oxygen and increases the carbon dioxide content of soil air.

Providing the covering vegetation is not too dense, sunlight penetrates the litter, open burrows, and superficial soil crevices. Below the surface, at depths of only a few centimetres at most, is permanent darkness. In deserts, grasslands, alpine regions, and deciduous forests after the leaves have fallen, the surface is largely exposed to the sun's radiation. Thus heated by the sun, the surface temperature of soil rapidly climbs well above that of either the air or the soil at slight depths beneath. Surface temperatures above 50°C are not infrequent even in the Temperate Zone. Owing to the low thermal conductivity of soil, the heating and cooling of deeper layers lags behind the fluctuations in temperature of the air, litter, and the soil surface. Heat acquired during day penetrates into the soil quite slowly. While the surface cools in the evening, the deeper layers of soil are still rising in temperature. In the morning, the deeper layers have cooled but the surface is now being heated by the sun, creating a steep, vertical gradient in temperature. A high moisture content in soil will reduce the amplitude of these daily fluctuations in temperature. "Cover objects," such as large rocks or logs resting on the surface, shelter the soil beneath from extreme temperatures and maintain a high humidity by preventing evaporation. The spaces beneath cover objects provide insects a place to escape from predators and environmental extremes, especially in grasslands.

Litter Insects. The most numerous insects in the litter and superficial layers of topsoil, in terms of both species and individuals, are colonies of Isoptera, colonies of Formicidae, all stages of Coleoptera, immature Diptera, and immature Lepidoptera, especially pupae. The most numerous arthropods of all here are the Acarina and Collembola. Each of the other insect orders is represented by at least one species, except the Ephemeroptera, Plecoptera, and Phthiraptera. The insects of certain orders spend their entire lives in litter, often in association with cover objects (Archeognatha, Thysanura, Grylloblattodea, Zoraptera, and some species of Blattodea, Dermaptera,

Embioptera, and Orthoptera). Other orders are represented at least by the immature stages which feed as scavengers or predators or which feed on algae, fungi, or mosses (Mantodea, Thysanoptera, Psocoptera, Neuroptera, Raphidioptera, Mecoptera, and Siphonaptera). Parasitoids and parasites, including Strepsiptera, attack other litter insects.

Remarkably few aquatic insects are also active in moist litter, even though the habitat would seem suitable and, in most instances, the aquatic forms probably evolved from litter inhabitants. Exceptions are the terrestrial naiads of Megalagrion (Odonata) in Hawaii, and the terrestrial larvae of Enoicyla pusilla (Trichoptera) in forest litter of Europe. Megaloptera and aquatic Coleoptera, however, regularly pupate out of water in soil. The transition between aquatic and terrestrial life is often made by fly larvae. For example, crane-fly larvae (Tipulidae) readily burrow from beneath streams into moist soil.

Immobile stages of insects are commonly spent on the ground. Eggs are often deposited here: Phasmatodea drop eggs in litter, and Acrididae lay eggs in soil. Lepidoptera pupae and the resting "pupae" of Thysanoptera are in litter. Periods of quiescence or diapause also may be spent among dead leaves. Aggregations of diapausing adults of Hippodamia convergens (Coccinellidae) are found on the ground in Sierra Nevada of California from June to February.

The active stages of litter insects are greatly varied in shape. Most are somewhat dorsoventrally compressed, and often parallel-sided or anteriorly narrow, a form that facilitates movement among varied obstacles. Most are agile and possess well-developed means for locomotion on ground. Escape by jumping is often a characteristic (Archeognatha, Collembola, Orthoptera, Schizopteridae (Hemiptera), alticine Chrysomelidae, and rhynchaenine Curculionidae). The antennae and tactile sense organs are usually well-developed. Exposed membranous wings are apparently an encumbrance in narrow passageways.

Except for mating and disperse flight is an unnecessary investment of energy and structure. The Embioptera spilt silken run ways to move out from cover objects into the litter. The wings of male web spinners, if present, are flexible and bend freely inside the tunnels. The Collembola, Diplura, Protura, and Apterygota are, of course, wingless, and the immature pterygotes have small wing pads at most. Among adult pterygotes some have protective forewings (Orthoptera, Dermaptera, and Caleoptera); others fail to develop wings (some Psocoptera, Grylloblattodea, and the sterile castes of Isoptera and Formicidae); or discard them after mating and dispersal (Zoraptera, reproductive castes of Isoptera and Formicidae).

Certain features of the body are evidently correlated with daily illumination and hazard of predation by visual hunters such as birds the compound eyes are well developed; circa lian rhythms in behaviour are present, often with nocturnal activity; and the body is often pigmented with concealing patterns. Litter insects are generally resistant to desiccation. Respiration is by tracheae, and the spiracles are usually fitted with closing devices.

Soil Insects. Below the litter layer and soil surface the variety of arthropod taxa drops abruptly, yet some groups are found here only. Collembola and Acarina are the most numerous of the permanent residents. All species of Pauropoda, Symphyla, Diplura, and Protura are entirely confined to soils, moist litter, or crevices under cover objects. No entire order of insects is similarly restricted, but some groups in various orders spend part or all of their lives underground.

The most common are Isoptera, Hymenoptera (Formicidae and certain other aculeates), Diptera, Coleoptera, Hemiptera, and Orthoptera. Of these, the Coleoptera are the most numerous in species. All stages of the life cycle of Coleoptera can be found in soil, but generally only the immature stages of Diptera are present. Termites and ants make their nests in soil. Their excavations and the plant matter brought into the nest cotribute to soil development. Galleries of Atta ants may be enormous and reach 3 to 6 m in depth. Many species of bees and predatory wasps excavate simple burrows in earth that are stocked with provisions for their young. Roots are the food of nymphal Cicadidae and certain other subterranean Hemiptera. Mole crickets feed on roots but also take insect larvae. These orthopterans and the omnivorous Stenopelmatinae (Gryllacrididae) are true inhabitants of the soil.

The high humidity of soil air, films of soil water, and moderate temperature of soil create conditions approaching an aquatic environment. Many invertebrate phyla are represented in soil by minute aquatic forms that are found elsewhere in freshwater. Protozoa, for example, are the most numerous soil animals. In the soil air of pore spaces live the smaller Collembola, Acarina, Pauropoda, Symphyla, Diploma, and Protura. When the number and size of organisms are compared to the surface area of the walls of pore spaces, it is evident that the cavities are greatly underpopulated (Christiansen, 1964).

An important factor limiting insects in soil is the obvious difficulty in moving about. Unlike the small arthropods listed above, insects are generally not small enough to negotiate interconnecting pore spaces for any distance. They must either tunnel by pushing aside particles and squeezing through, or excavate the soil in front and deposit it behind. The physical resistance of the soil, therefore, raises the energetic cost of locomotion and reduces the distance to which an individual can hunt for mates or food. Sight and wind-borne odours are also eliminated as aids in the search. For this reason, mating of insects is usually accomplished above ground. Except for the roots of green plants, insects must find their food in the form of other soil organisms, dead or alive Active excavators such as ants, termites, silphid beetles, larval cicindelids, warps, and bees obtain their food from richer sources above ground.

Reduction in wings is a characteristic of soil insects as well as of litter insects. Soil insects are generally round in cross section but are greatly varied in shape otherwise. "Tunnelers" have reduced or no legs and are of two kinds. Larval Elateridae and Therevidae are smooth, stiff-bodied and slender. By sinuous movements of the trunt they force their hard, tapered heads forward through the substrate. Larval Tipulidae and Bibionidae are soft-bodied and use peristaltic movements to penetrate soil. "Excavators" may be fairly thick-bodied and often exhibit conspicuous molel. A modifications of the head and forelegs for digging. Such fossorial adaptations are seen in Gryllotalpidae, Cydnidae, nymphal Cicadidae, and Searabaeide (Copris, Geotrupes), to list only a few. Worker ants and termites are not so clearly equipped for digging, but they effectively use their mandibles to remove and carry away particles.

In the absence of light soil insects and the other hexapodous arthropods send to have reduced or no compound eyes, antennae variably developed, even absent (Protura); well-developed partile sense organs and pale pigmentation or the integument without patterns. They are usually repelled by light and higher temperatures. The upper limits of tolerance to temperature are relatively low, but some soil insects can remain active a few degrees above 0°C. Rhythmic behaviour has been poorly studied. Vertical migration in response to daily temperature fluctuations may occur if the soil exposed to solar radiation. Resistance to desiccation is low, probably because the epicuticle becomes abraded by soil particles or is naturally permeable to water. Some respire cutaneously. Those that normally inhabit the deeper subsoil are tolerant of higher levels of carbon dioxide, at least for several days (some Collembola, larval Elateridae, Scarabaeidae).

Many insects that inhabit deserts are frequently dependent on the sandy substrate for protection againts extreme temperatures and desiccation. Those running on the surface during the day may have long legs that lift the body well above the hot surface (Mutillidae, Tenebrionidae). During the hot season many forage openly only at night, spending the day in burrows or simply immersed in sand. Edney (1974) has shown that surface temperatures up to 72°C and humidity of 10 per cent are avoided by the desert cockroach Arenieaga investigata. At depths of 30 to 45 cm in sand, the roach inhabits a zone where daily temperatures are well below the upper lethal limit (about 45.5 to 48,5°C) and humidity may reach above 82 per cent, a level at which the roach can actually absorb water vapour. Like some tenebrionic beetles, the roach literally swims in sand. The body is biconvex in cross seetion, oval, and smooth and has sharp lateral margins, resembling that of an aquatic beetle.

Cave Insects. Insects found in caves may be the same species that are normally found in litter or soil outside. Certain flying insects, especially flies (Culicidae, Mycetophilidae), seek shelter in caves during the day. The darkness, high humidity, even temperature, and absence of green plants are conditions similar to those of deeper soil, but without the physical restraint. Soil insects, therefore, already have many of the modifications required for continuous life in caves.

The absence of plants except for their roots, however, largely limits cave-dwelling insects to scavenging, parasitism, or predation. Phytophagous insects are usually absent. New organic matter must be washed in or be deposited as waste by organisms that enter and leave. Bats attract several taxa of parasites (Polyctenidae, Streblidae, and Nycteribiidae; some Siphonaptera and Cimicidae) and the accumulated feces of bats (called guano) attracts scavengers (Arixeniina, Dermaptera; Blattodea, Diptera, etc.).

Animals which are adapted to continuous life in caves and which reproduce there are called troglobites. Insects and other hexapodous arthropods in this category are almost entirely limited to Collembola (Entomobryidae), Diplura (Campodeidae), Orthoptera and Coleoptera (Pselaphidae; trechine Carabidae, and additionally in Europe, bathysciine Leiodidae). Aquatic insects are absent, but some cave Collembola and beetles have been observed submerged and alive in bodies of water in caves.

AQUATIC INSECTS

In this section we discuss insects that spend at least the immature stages of their lives floating or submerged in water.

Aquatic insects are a minor fraction of all insects, probably numbering no more than 3 to 5 per cent of all species. The limited number is probably the result of the limited amount of freshwater habitat in comparison with the land surface. Yet these insects are taxonomically diverse (Table 12.1), are fascinating in structure and biology, and some of them, such as mosquitoes, are of extreme importance in public health.

A few insects are truly marine. Some others live in the intertidal zone, in brackish coastal waters or saline lakes. Most aquatic insects, however, are restricted to freshwater, where they are among the most important organisms in the aquatic ecosystem. As herbivores, predators, scavengers, and even parasitoids, insects function in the trophic structure and, in turn, are food for fish, birds, and amphibians.

Aquatic insects exhibit conspicuous adaptations for respiratoin and locomotion within the constraints of a liquid medium. Similar adaptations, especially for respiration, can be observed among insects in semiliquid media: scavengers in decaying plant or animal bodies or animal waste; phytophagous insects that burrow in juicy fruits or other plant tissues; and endophagous parasites and parasitoids immersed in the hosts' blood.

Extent of Freshwater Habitats. On a global scale the total extent of freshwater is remarkably small. Freshwater lakes and rivers constitute less than 0.01 per cent of the total volume of water in the biosphere) Wetzel, 1975). Even including the inland seas such as the Black and Caspian, all inland bodies of still water occupy only 1.8 per cent of the land surface. Rivers or running water occupy about 0.001 per cent. Much freshwater is concentrated in a few large lakes in the Temperate and Subarctic regions of the Northern Hemisphere. Lake Baikal in Siberia and the Laurentian Great Lakes of North America each contain about 20 per cent of all freshwater.

Freshwater has been available to terrestrial life because of the hydrological cycle. Actual geological evidence of ancient freshwater habitats is scarce. The sediments of lakes and rivers, like other terrestrial deposits, seldom escape destruction by erosion. At any instant lakes contain about 100 times more water than rivers, but rivers are continually renewed. Over a year rivers carry a volume equal to perhaps a third of all freshwater lakes (Hynes, 1970a).

Insects of Running Water. Although the total volume of running water is less than still water, continental drainages exist for periods of a million years or more. This provides a continuously habitable environment for aquatic organisms over a wide geographic area. Because major drainages are more or less permanent and are otherwise favourable, insects have evolved the greatest variety of taxa in running waters. Indeed most kinds of aquatic insects are represented in running waters, and some occur only there. This is especially true in the rhithron, or rapidly flowing, shallow, rocky streams of the headwaters where oxygen concentration is high and the mean monthly temperature seldom exceeds20°C.

Water seeping into a stream from the ground or issuing from springs is commonly cooler than air in summer and warmer than air in winter. This difference may be sustained for some distance if the stream is shaded in summer or insulated by a snow cover in winter. Although subject to daily and seasonal fluctuations, the range of variation is usually less than that of the shallow regions of lakes. Furthermore, cooler water absorbs much more oxygen than warmer water, *e.g.*, water saturated at 4°C holds 1.5 times the amount at 225°C. Turbulence over the rocky bed maintains dissolved oxygen at the point of saturation and distributes it to the depth of the stream. Consequently, insects are able to feed and grow throughout the year, even in winter in Temperate regions.

Spec es of Ephemeroptera, Plecoptera, Trichoptera, and Megaloptera are the most numerous in streams. Larvae of the fly families Blephariceridae, Simuliidae, and Deuterophlebiidae, and the beetles Helminthidae, Dryopidae, Elmidae, Ptilodactylidae, and Psephenidae occur among rocks in swift, cool waters. Naiads of the odonates Agrionidae, Cordulegastridae, and Gomphidae also favour running water.

An ever-present hazard to current dwellers is to be swept off the bottom and drift downstream exposed to predators. The number of insects drifting usually rises on darker nights, because many are nocturnally active. Drifting may involve only short distances and not be entirely deleterious. Competition for space is relieved, and the population is more evenly distributed downstream. Evidence is inconclusive that adults tend to fly upstream to lay their eggs and recolonize the uppermost waters. The eggs of mayflies, stone flies, and some dragonflies are equipped to sink rapidly, and firmly attach to the bottom, thus remaining near the ovipositional site.

Insects that inhabit swift water are generally prevented from visiting the surface for air. Their respiratory structures are tracheal gills or plastrons that function while bathed in oxygen-rich water. Naiads of some mayflies and stone flies that normally reside in still pools create their own currents. The abdominal gills of mayflies have an undulating beat that circulates the surrounding water. Similarly, stone flies pump up and down on their legs at a rate proportional to the oxygen need. Insects living in the swiftest waters do not exhibit such respiratory movements and depend entirely on the current to bring oxygen. They die quickly even when placed in still water of the same temperature and oxygen saturation because the oxygen supply becomes depleted in the proximity of the gills and is not renewed by currents.

A number of similar morphological adaptations has been evolved independently by various insects in torrential waters (Hynes, 1970a). The body may be dorsoventrally compressed so that the insect can creep in the thin boundary layer of 2 to 3 mm next to the substrate where the current declines to zero. Naiads of the mayflies Heptageniidae, and larval beetles of Psephenidae are thus flattened. Small size also permits an insect to utilize the boundary layer. Nearly all adult beetles in swift water are small.

Various attachment devices allow insects to cling to the substrate. The beetles Elmidae and Dryopidae have the last tarsal segment and claws enlarged for anchoring themselves. In addition, the larvae of Elmidae have prehensile hooks near the anus. Stout tarsal claws and anal hooks also are found on the free-living caddis larvae of Rhyacophilidae and the hell-grammites or larvae of Corydalidae (Megaloptera). Lacking true legs, the aquatic larvae of the fly Families Simuliidae, Deuterophlebiidae, Blephariceridae, Chironomidae, Rhagionidae, Empididae, Syrphidae, Ephydridae, and Limnophera (Anthomyiidae) have developed leglike prolegs for attachment. The prolegs are usually fitted with fine hooks. Suckers are developed near the anus of larval simuliids and on the ventral surface of Maruina (Psychodidae) and Blephanceridae. Frictional resistance to detachment is developed by ventral gills or hairy pads on the mayfly naiads Rhithrogena (Heptageniidae) Ephemerella doddsi and E. pelosa (Ephemereliidae) the African Dicercomyzan (Tricorythidae), and the larval beetles of Psephenidae. Silk and sticky secretions are used for attachment by some Chironomidae, and the caterpillars of Pyralidae (Lepidoptera). Larval simuliids move about with their prolegs on silken mats. Hydropsyche (Trichoptera) spin nets in the current and harvest waterborne food. Most Trichoptera attach their pupal cases to rocks with silk. The stone cases built by Trichoptera in streams are usually composed of heavier grains and a few heavy pebbles. This provides ballast so that the caddis larva quickly sinks into still water if dislodged.

Few insects actively swim against the current. The streamlined, minnowlike mayflies Ameletus, Isonychia, Centroptilus, and some Baetis move in and out of the current with powerful, up-and-down movements of their caudal filaments. On the surface of riffles are the bugs Rhagovelia.

Insects in crevices can utilize the rich supply of oxygen and food of streams without the immediate danger of the current. Flattened mayflies are also found in the dead water beneath rocks. Tiny, young instars of insects, the narrow-bodied stone flies and mayflies, and crane-fly larvae, Antocha and Hexatoma

(Tipulidae), normally are found amongst the bottom gravel. Leaves packed against obstructions have a special fauna of mayflies, caddis flies, and stone flies. The stout-limbed Ephemerella is commonly found here.

The lower region of a drainage, or potamon, is characterized by slower flow, greater depth, occasional low oxygen concentrations, and monthly mean temperatures close to that of air and rising above 20°C. The bed of the river is formed by deposits of sand and mud.

Fewer kinds of insects are specialized for life in the potamon, but some species are extremely abundant. Naiads of the burrowing mayflies Ephemeridae excavate U-shaped burrows in silt. Undulating movements of their feathery gills circulate water in the burrow, bringing in organic matter, which is eaten, and oxygen (Fremling, 1960). Other mayflies, such as Caenidae, Tricorythidae, and some Ephemerellidae, inhabit silty bottoms of streams and rivers. These have the functional gills kept clean of particles by the operculate uppermost gill. The body surfaces of some bottom-dwelling mayflies, stone flies, and dragonflies are covered by a protective layer of fine hairs. Larvae of chironomid midges are often abundant in soft, mucky bottoms where they build tubes.

Insects of Still Waters. Lakes are relatively short-lived in comparison to rivers. Most lakes existing today originated in the Pleistocene and are less than 25,000 years old. Only a few large, deep lakes are much older. Lake Baikal and Lake Tanganyika in Africa probably existed in the Tertiary Bodies of still water normally fill with sediments or evaporate during prolonged dry periods.

In contrast to running water, the insect life of still water is influenced more by temperature and oxygen supply than by water movement. Temperatures in shallow, marginal waters of lakes or small bodies of water such as ponds tend to fluctuate with air temperature. Large bodies of still water warm slowly in summer, becoming stratified into an upper, warmer layer, or epilimnion; a zone of transition, or thermocline, where temperature decreases 1°C or more per meter of depth; and a deep layer at 4°C called the hypolimnion. In winter, lakes cool slowly. Water reaches maximum density at 4°C and is less dense at lower or higher temperatures. Surface water cooled to 4°C sinks. Ice forms at the surface after deeper water is uniformity 4°C and the colder, lighter water is at the top. Insects in lakes are thus protected against freezing. Oxygen is absorbed at the water's surface and is produced by photosynthetic plants. In summer the oxygen supply of the epilimnion may exceed saturation. Decomposition in the hypolimnion reduces the supply.

Lakes that are poor in nutrients are called oligotrophic. Young, deep lakes are characteristically of this type. Precipitation and runoff from the watershed carry dissolved substances into the lake, promoting the growth of free-floating algae in sunny, open water. The gradual enrichment of water with nutrients, especially nitrogen and phosphorus, is called eutrophication. Continued addition leads to the production of organic matter in excess of decomposition. The decomposed organisms in deep water deplete the water of its oxygen, leaving the organic matter to accumulate with other sediments. Older lakes, rich in nutrients and shallow, are called eutrophic, When lakes receive large amounts of acid humic substances from the watershed, the water becomes stained brown and poor in variety of aquatic life. Organic matter accumulates from plants in the shallow marginal waters. Such lakes or bogs are called dystrophic. Ultimately, both eutrophic and dystrophic lakes fill completely with organic matter, are overgrown by vegetation, and disappear.

Slender fly larvae dominate the insect life of the deeper, open waters and the bottom beneath. The thin sediments of oligotrophic lakes are characterized by colourless midge larvae of Tanytarsus (Chironomidae). On the surface of the bottom ooze of eutrophic lakes, where oxygen is often depleted, the predominate or only insects are red, hemoglobin-containing larvae of Chironomus. These are able to survive the irregular oxygen supply and continue to function as important decomposers. The open waters of eutrophic lakes are inhabited by the phantom midge larvae, Chaoborus (Culicidae). As one of the few free-floating insects, the predatory Chaoborus spend the day near the bottom and swim toward the surface at night. The phantom larva is nearly transparent—hence the name. Gas sacs are visible at each end and act as organs of equilibrium. The volume of gas is controlled by the nervous system. When the sacs are compressed, the larva sinks: when they are expanded, it rises.

Adult midges emerge synchronously in enormous numbers from productive lakes throughout the world. Chaoborus edulis forms spectacular clouds seen at great distances over Lake Malawi in Africa. The emergence of C. astictopus, the Clear Lake gnat of Clear Lake, California, creates an annual nuisance to resort owners. When carried over land by winds, the masses of chironomids and Chaoborus remove significant quantities of organic matter from the lake ecosytem.

The shallow water of lakes and ponds contains more different kinds of insects than the deeper, open waters. Certain areas approximate the physical and chemical conditions of running water. On the wave-washed, rocky shore of oligotrophic lakes we find some of the same mayflies, stone flies, and caddis flies that are in streams. Stone flies are sensitive to low oxygen and are rare in lakes except in such places. The numbers and variety of insects in shallow water increase with eutrophication. Mayflies, caddis flies, dragonflies, and damselflies live among submerged plants or on the bottom. These insects have gills and do not rise to the surface for air. Also confined to quiet water are the many Hemiptera, Coleoptera, and Diptera that breath air, either directly or in air bubbles. On the surface are the water striders, Gerridae, Veliidae, and marsh treaders, Hydrometridae.

Shallow lakes may be created each year by rain and snow melt. These last but a few weeks or months before the water is lost through evaporation or absorption into the soil. Such bodies of water formed in the spring are called vernal lakes or ponds. Insect life ordinarily teems in these temporary habitats because predatory fish are absent. Even the small volumes held in tree holes, cavities at the bases of leaves of plants such as the tropical bromeliads, and hoofprints are sufficient for some insects to complete their life. Growth is usually quite rapid. For example, less than 2 weeks is needed for the yellow-fever mosquito, Aedes aegypti, to complete development.

Insect Scavengers

Many, but by no means all, insects that live in soil and water are scavengers. Perhaps the word "scavenger" is poorly chosen. Does it indicate an animal that eats whatever comes along, one that eats mostly dead things, or one whose diet consists of items so small and miscellaneous that an accounting is difficult? Here we intend all these meanings when we call certain insects scavengers. The term is broad enough to convey the diversity of food that some insects are known to eat while also concealing our ignorance of the exact diets of others.

A complex terminology has developed to describe the diets of scavengers. Some may be omnivorous, eating plants or animals whether they are dead or alive. Saprophagous scavengers feed on dead organisms and have been subdivided into xylophagous forms, boring into and feeeding on sound or decaying wood; phytosaprophagous formes, feeding on decaying vegetable matter; scatophagous or coprophagous forms, feeding on feces or dung; and zoosaprophagous or necrophagous forms, feeding on dead animals. Detritivores feed on small bits of animal or vegetable trash. The food of saprophagous insects, however, is not all dead matter, because microbes such as bacteria, fungi, and yeast flourish in such places and are nutritious.

Small living food is taken by mirophytic insects that selectively eat bacteria, yeasts, fungi, algae, diatoms, lichens, spores, and loose pollen. Animal foods include Protozoa, and small invertebrates and their eggs. Fungivorous insects are adapted to feeding exclusively on living and dead fungi. Thus among the insects loosely called scavengers are those with specific choices of food. Major taxa of scavenging and fungivorous insects are listed.

Saprophagous insects have unity in an ecological sense because they function in the complex world of the decomposers. These should be counted among our most beneficial insects. Intact bodies of dead vascular plants and vertebrates, smaller dead organisms, and excrement are progressively disintegrated. The nutrients in streams and soils are deprived from plant material that is decomposed initially by insects and other arthropods. Scavenging species are usually specific to either plant or animal matter. In terrestrial environments the species appear in a predictable sequence as characteristic chemicals are emitted from the decaying organism at each stage of disintegration. Tissue is ingested and partially or completely digested, often with the aid of symbionts or unusual enzymes such as cellulase, keratinase, or collagenase. The feces of scavengers are further decomposed by other scavengers or microbes, thus aiding in nutrient recycling.

Early Evolution of Insects in Soil and Water

It is generally agreed that insects evolved from some form of marine arthropod. Opinions vary, however, as to whether the intermediate steps in this evolution took place in water or on land. In other words, were the first insects originally aquatic and secondarily terrestrial or vice versa? The weight of evidence favours a terrestrial origin. The tracheal system of insects and their allies seems to have evolved for the transport of air in a terrestrial environment. The aquatic stages of insects possess the same gas-filled tracheae as terrestrial insects. Only by special modifications can the tracheal system function under water. Even then, many insects must return frequently to the surface for air.

In most groups of aquatic insects the immature stages alone are aquatic. The adults are aerial or terrestrial, returning near the water or reentering it for oviposition. The aquatic Hemiptera and Coleoptera are the only large groups of insects that live submerged in water as adults. Neither has entirely relinquished its terrestrial adaptations: both usually retain functional wings, and most beetles pupate in soil away from the water. Furthermore, only the specialized Aphelocheiridae (Hemiptera) of Europe, and possibly some Capnia spp. (Nemouridae, Plecoptera) are able to remain submerged throughout their lives. For these reasons, one must conclude that insects were originally terrestrial and are incompletely adapted to life under water.

During the evolutionary history of insects, aquatic forms have evolved repeatedly, perhaps hundreds of times, from terrestrial ancestors. Often, but not always, the more primitive orders of a group of orders, or the more primitive families in an order, are aquatic. Life in water offers certain advantages, especially for the immature stages. They are sheltered from desiccation and freezing and protected from certain natural enemies such as parasitoids. The aquatic habitat was apparently colonized early during each new evolutionary radiation of insects, and the aquatic forms tended to outlast their terrestrial progenitors. Hence the more primitive forms are well represented in the aquatic enviroment because they survived there, not because insects originated in water.

Turning now to the early terrestrial environment, the first episode of plant evolution on land began in the late Silurian Period and ended with the close of the Devonian Period. During this relatively short span of time, a rapid succession of vascular plants appeared: the Subdivisions Psilopsida, Lycopsida, Sphenopsida, and the ferns or Pteropsida. The diversity of this early flora can be estimated by the 83 "genera" of spores that have been described (Chaloner, 1970). When the sporophyte plant was preserved, it was sometimes treelike. The Devonian Pseudosporochnus reached 2 m or more in height. All these plants were confined to wet places because the flagellated sperm of the gametophyte generation required water in which to swim tó the egg cells.

The fossil record indicates that a variety of aquatic mollusks, worms, and arthropods could have served as decomposers when plants fell in the water. But on land, few animal fossils exist to reconstruct the ancient food webs of the early Paleozoic Era. The moist, decaying plant matter would have been a suitable place for arthropods to make the transition from aquatic to terrestrial life. Here was an uncontested food supply, including spores and microbial life as well as protection from desiccation.

The fossils that do exist from this time support this line of reasoning. Predatory scorpions are represented in both the Silurian and Devonian Periods. The earlier forms may have been aquatic, but the later scorpions were terrestrial. A myriapod, Archidesmus, is known from the Silurian and may have been a terrestrial scavenger. In the Devonian a mite, Protacarus, and a few small arachnids have been found. The mites were probably microphytic and the arachnids predatory. Thus some evidence exists that arthropod scavengers and predators colonized the land as early as the Silurian and Devonian. Furthermore, they probably lived in litter. Certain protective characteristics of the spores of terrestrial plants and fungi that lived at this time indicate that they may have been fed upon and dispersed by arthropods. Fragments of arthropods have also been found inside a fossil stem fragment and inside fossil sporangia, further supporting the idea that arthropods probably thrived in litter (Kevan *et al.*, 1975).

Definite insect fossils are not known from the Devonian. yet about 30 million years later an abundance of varied insects are found in strata of the Upper Carboniferous Period. It seems most probable that insects evolved as scavengers in plant litter and made their appearance at least by the close of the Devonian. The Apterygota are probably representative of the early insects, and it is not coincidental that they are scavengers in litter today. Other survivors of the same early radiation of terrestrial arthropods include the Pauropoda, Symphyla, Collembola, Diplura, Protura, Chilopoda, Diplopoda, and the arachnids Pseudoscorpionida, Scorpionida, Palpigrada, Podogona (Ricinulei), and the more primitive Acarina and Phalangida. These groups are also found today mostly in moist soil and litter.

Evolution of Scavengers. As the diet of Paleozoic insects became predominantly living prey, blood, fungi, or green plants, scavengers became predators, parasites, fungivores, or herbivores respectively. Most evolutionary lineages in insects seem to be form scavenging to other feeding habits. Except for the mouthparts of larval and adult Diptera, the mouthparts of the scavenging taxa are of the chewing type. This versatile design is readily modified into more specialized feeding structures.

In the Carboniferous Period, the amount of food for scavengers increased enormously. A rise in the carbon dioxide content of the atmosphere is thought to have created a "greenhouse effect" and produced a warm, moist climate. The increased availability of carbon dioxide plus a favourable climate resulted in an increase in photosynthetic activity and the explosive growth of land plants. The dominant lycopod, Lepidodendron, reached over 40 m in height. The rapid accumulation of dead plants created the great coal deposits upon which we now depend for fuel. The animal decomposers preserved as fossils were insects, mites, millipeds, and giant arthropods (now extinct) of the Class Arthropleurida. The last were shaped like millipeds and measured up to 2 m in length. They were probably the world's largest exclusively land arthopods and a dramatic measure of the amount of food available to scavengers. Other decomposers, such as pulmonate snails and terrestrial isoponds, were apparently absent.

At this time insects evolved wings and the Paleoptera made their appearance. Adult Paleoptera, with their outstratched wings, would have been prevented from returning to life among the crevices in the litter. The nymphs were probably less restricted, and those with chewing mouthparts probably continued as scavengers. The scavengers and herbivores were prey to increasing numbers of predators. These now included amphibians and the first reptiles, as well as the first spiders and other arachnids, predatory insects, and possibly chilopods. The pressure to escape terrestrial predation may be responsible for the origins of the aquatic nymphs in the Odonata and Ephemeroptera. The latter continued their scavenging habits in their new environment.

The evolution of the wing-folding mechanism permitted the entire life cycle of the early Neoptera to be completed in or near the litter. The success of this adaptation can be judged by the prevalence of scavengers among the early Neoptera that survive today Furthermore, the first scavenging Neoptera to be fossilized were the Blattodea. They were so abundant that the period is often called the "Age of Cockroaches."

The Permian Period was characterized by geologic and climatic disturbances, often violent. Some areas were glaciated, some became wind-blown deserts, and in other regions thick salt deposits were laid down from drying seas. One can imagine the stresses placed on the Carboniferous insect fauna that had become adapted to moist, warm forests over a period of 80 million years. Besides struggling to survive under new climatic regimes and almidst the competition caused by shrinking forests, they were confronted with an increasing variety of reptiles which doubtless included insectivores.

During this period of changing and adverse conditions, the holometabolous life cycle was evolved. The larvae of scavengers were able to bore into decaying tissues, or into soil, thus gaining increased protection from desccation and predators, as well as access to new food sources. Among the first scavenging endoptergotes to be preserved as fossils are Coleoptera of the Family Cupesidae. This family survives today, and the larvae bore in decaying wood.

Through the Mesozoic and into the Tertiary, the scavenging endopterygotes became specialists on larger, dead organisms and their wastes. Wood and decaying foliage of the gymnosperms and angiosperms were penetrated. Alone among the exopterygote orders, the Isoptera were also able to penetrate sound wood. The dead bodies of reptiles and later mammals, and their excrement, were invaded by insects able to feed and respire in semiliquid media. The higher metabolic activity of mammals was accompained by higher food intake and increased quantities of feces. Dead animals and mammalian feces as habitats are exclusively occupied by endopterygotes, especially Coleoptera and Diptera.

Scavengers in the nests of birds and mammals became scavenging ectoparasites such as the Phthiraptera (Suborder Mallophaga), and some became obligate bloodsuckers. Certain dung-feeding maggots myiasis-producing parasites. Insects feeding on seeds stored by mammals or the provisions of presocial and eusocial insects later became pests of human granaries and pantries.

10

Inter-Relationship

Insects are the vectors as each one carries with it a small community of other organisms. Some are merely phoretic passengers, but others live more or less intimately in or outside their hosts. In the high mountains of New Guinea, large fightless weevils live amidst mosses in humid forests. The backs of the long-lived, slow-moving weevils are camouflaged with living fungi, algae, lichens, and liverworts. The integumental surface and setae, plus a secretion, encourage plant growth. Mites, nematodes, rotifers, diatoms, and psocid insects live and feed among the plants, causing no harm to the weevils.

Several microbes and occasional helminth worms are usually associated with all other insects. Lower forms of life were psesent before insects evolved. We can speculate that as scavengers in decaying vegetable matter the early insects would have been immersed in a microbe-rich environment and would have ingested and excreted saprophytic forms. Inside the insect's gut, some were digested and absorbed, and some persisted harmlessly or as injurious parasites. In time, some microbes evolved a beneficial partnership such that their hosts could exploit foods that were otherwise nutritionally deficient. The mobility of insects and their resistance to desiccation allow their less hardy associates to survive and be dispersed in the terrestrial environment. When insects began feeding on the vascular fluids of higher plants and vertebrates, certain viruses and microbes were introduced into the wounds. Some proved to be highly dangerous to their new hosts. Insects also became involved as intermediaries in the natural cycles of other disease agents. Although insects alone rarely cause death of plants and vertebrates, as vectors of disease they can have catastrophic effects on entire populations.

Symbiotic Relationships

Broadly defined, symbiosis is the living together of dissimilar organisms, regardless of the possibly injurious or beneficial interactions. In our present discussion, the microbes are called symbionts and insects are the hosts.

Symbionts may live parasitically at the host's expense. If the host's vitality is impaired, the symbiont is called a pathogen. Agents causing fatal diseases of insects include certain viruses, bacteria, protozoa, fungi, and probably most parasitic helminths. Some of these pathogens can be used effectively in controlling pest insects. Insect hemocytes defensively phagocytize the smaller agents and encapsulate the larger ones. Insects do not produce specific antibodies against pathogens, but lysozymes in the gut, fat body, and hemolymph can destroy microbes by enzymatic action.

A coevolutionary tendency exists for the hosts and symbionts to become mutually adjusted. Parasites that kill their host also die, whereas those causing less harm to their host survive. Hosts that are susceptible to parasitic damage are selectively eliminated, leaving resistant hosts in the majority. Thus the relationship evolves toward benign parasites and resistant hosts.

Symbionts that cause no harm to their hosts are commensals, such as the fungi Laboulbeniales. Mutualistic symbionts have evolved a reciprocally beneficial relationship with their host. When the symbionts are outside the insect's body, *e.g.*, the fungus gardens of insects, they are called ectosymbionts. Endosymbionts are mutualistic microbes sheltered inside the insect's body. Buchner recognized that certain groups of insects were dependent on symbionts because the insects' diet lacked essential items. For example, the vascular fluids of both higher plants and vertebrates, and wood are deficient as food for insects.

Phloem-sucking Homoptera obtain carbohydrate in surplus, but the diet is inadequate in protein; all possess symbionts that supply missing nutrients. Predatory Hemiptera lack symbionts but phytophagous Pentatomidae have them Larvae of Dasyhelea flies (Ceratopogonidae) and Nosodendron beetles (Nosodendridae) live in sap and have symbionts.

Vertebrate blood is notably lacking in B vitamins. Phthiraptera, Cimicidae, Triatoma (Reduviidae), Glossina (Muscidae), and crawling ectoparasitic flies are parasites that feed exclusively on blood in all feeding stages and have symbionts. Adult parasites with scavenging larvae, *i.e.*, Siphonaptera and biting flies, lack symbionts. The larvae of these insects obtain B vitamins sufficient for the rest of the life cycle from bacteria-rich food.

Wood-feeding insects depend on ectosymbionts such as the fungus gardens mentioned above or on endosymbionts in the intestine. Scarab grubs contain bacteria; anobiids and certain cerambycids contain yeasts; and certain termites contain protozoa or bacteria. Certain pests of seeds and stored grain have symbionts: Nysius (Lygaeidae), Lasioderma (Anobiidae), Sitophilus (Curculionidae), Rhizopertha (Bostrychidae), Oryzaephilus (Cucujidae), and Coccotrypes (Scolytidae). Oddly, the Bruchidae apparently lack symbionts.

Symbionts also occur among insects with a variety of other food habits. Many phytophagous weevils harbour bacteria. Bromius, Casside and Donacia are chrysomelid beerles whose larvae feed on green plants and have bacterial symbionts; other chrysomelids lack them. Tephritid flies and lagriine tenebrionid beetles feed in fresh or decayed plants. Both have bacterial symbionts. Many species of scavenging cockroaches have bacterial symbionts, as do certain ants of the genera Componotus and Formica. The yeasts of nonpredatory Chrysopa adults have been described previously.

Microbes commonly gain entrance to the insect's body through the mouth as a part of the food. It is not surprising to find most endosymbionts still reside in the gut, having evolved the ability to live in the intestinal environment. They are often transmitted from generation to generation, or among individuals simply by ingestion. For example, the cellulose-digesting protozoa and bacteria of termites are passed among the colony members by anal feeding. Insects without overlapping generations infeet the eggs. Phytophagous Pentatomidae defecate on the eggs and the nymdhs acquire symbionts after hatching. Some insects have special organs for smearing bacteria on the egg as it is deposited. The female Donacia covers the egg with a secretion that encloses a mass of bacteria near the larva's head. On hatching. the larva eats through the mass and ingests the bacteria.

The host's intestine may be modified to accommodate endosymbionts. The hindgut of certain scarab beetle larvae is greatly enlarged as a fermentation chamber containing becteria and wood particles. Lateral saclike chambers, or caeco, are characteristic of insects with intestinal endosymbionts. Cranefly larvae may have a large diverticulum on the hindgut. Phytophagous Hemiptera-Heteroptera characteristically have many bacteria-filled caeca along the posterior portion of the hindgut. Malpighian tubules may also serve as bacterial crypts.

Insects as Vectors of Disease

Among the most devastating diseases of crop plants, domestic animals, and humans are those carried by insects. Insects that transmit disease agents are called disease vectors. The life history, host selection, and feeding behaviour of the vector largely determine the epidemiology of the disease. By controlling the insect, it is sometimes possible to break the link in transmission and to stop the spread of disease.

Microbes causing disease in plants and animals rarely also cause disease in the vector even when they live and multiply in the vector. An exception is Western X mycoplasma, which reduces the longevity of its leafhopper vector. Yet certain agents of plant disease actually seem to be beneficial to the vector by favourably changing the host plant's structure or functioning. Experiments have shown that vectors may feed or ovipost more frequently on diseased plants. A comparable relationship between animal diseases and their vectors is not known.

The nature of the transmission of pathogenic organisms by insect vectors to plant hosts may be mechanical or biological. Mechanical transmission involves the contamination of the body or mouthparts, or at most the foregut, after feeding on an infected host. The pathogen is transferred to a new host when the vector next feeds or makes contact. Infectivity is lost when the supply of pathogen is depleted or at ecdysis if the vector is immature. For example, mosquitoes function literally as flying pins in the transmission of the myxoma virus among rabbits. Similarly, the nonpersistent plant viruses are rapidly spread by aphids during their brief and repeated probing behaviour in search of suitable plant hosts.

Biologically transmitted pathogens, acquired after prolonged feeding. enter the vector's gut, hemocoel, or various organs. Infectivity is not immediate, but occurs only after a latent period during which the pathogen moves about in the host or further develops or reproduces. Once infective, the vector tends to remain infective for life.

The persistent circulative plant viruses, such as the aphid-transmitted pea enation mosaic, do not replicate in the vector, but circulate in the body. On the other hand, helminths causing animal diseases undergo developmental changes but do not multiply in their intermediate intect hosts. This is called cyclodevelopmental transmission of animal pathogens.

When the pathogen multiplies in the vector, the transmission is called propagative. Persistent propagative plant pathogens, such as the aphid-transmitted lettuce necrotic yellows, and plague bacteria in fleas, are examples. If the multiplication is accompanied by cyclic development, as in the case of malaria plasmodia in anopheline mosquitoes, the term is cyclopropagative.

Certain insects and acarines acquire infections as immatures and remain infective after one or more ecdyses. This is called transstadial transomission. Some transmit pathogens to their offspring by transvarial transmission, a feature already described above for endosymbionts. Certain arboviruses (see below) and rickettsiae are transovarially transmitted by ticks, but few insects have been shown to transmit animal pathogens similarly to their offspring. Plant pathogens, however, can be transsovarially transmitted. Some persistent viruses, rickettsialike pathogens; and mycoplasmas are transmitted in this way by leafhoppers.

INSECTS IN RELATION TO VIRUSES AND MYCOPLASMALIKE DISEASE AGENTS

Viruses are highly variable, submicroscopic particles consisting of either deoxyribonucleic acid (DNA) or ribonucleic acid (RNA) and usually, but not always, with a protein cover. Viruses multiply or replicate only in living cells, primarily those of bacteria, plants, insects, and vertebrates. The best-known viruses associated with insects are those that are pathogenic to insects, higher plants, or vertebrates. David (1975) and Vaughn (1974) have reviewed the viruses affecting insects, and Falcon (1976) discusses their use in controlling pests. From the standpoint of numbers of species of insects, the most important pathogens are the DNA-containing nuclear polyhedrosis viruses (NPV), DNA granulosis viruses (GV), and RNA cytoplasmic polyhedrosis viruses (CPV). Lepidoptera catterpillars and larval sawflies are most susceptible to these viruses. The polyhedrosis viruses are characterized by polyhedral crystals.

Carter (1973) lists 261 viruses that are transmitted to plants by insect vectors. Aphids transmit the largest number of viruses, mostly those producing the mottled "mosaic" diseases. Aphids also rank first in numbers of species of vectors. Some species may vector numerous viruses, for example, Mazus persicae is the vector of more than 100 viruses.

Leafhoppers rank next in importance and are noted as vectors of persistent propagative viruses. Recently, one group of diseases called the "yellows diseases" has been discovered to be caused not by viruses, but by organisms called mycoplasmas in the phloem that resemble minute bacteria (Whitcomb, 1973; Whitcomb and Davis, 1970). Like bacteria, mycoplasmas are procaryotes, *i.e.*, the nucleus lacks a membrane. The cells are highly variable in shape and uniquely lack a cell wall. Mycoplasmas were previously known as pathogens of mammals and birds, but not associated with arthropods. It has been estimated that as many as 350 species of plants may be involved in the natural cycle of aster yellows disease and its vector, the aster leafhopper, Macrosteles fuscifrons (Cicadellidae).

Of historical interest is the tulip "breaking" virus that causes the petals of red or purple flowers to be beautifully striped. The first record was published in 1576, making this the oldest known plant disease (McKay and Warner, 1933). Broken tulips became greatly admired is Europe. Paintings of the sixteenth and seventeenth centuries often depict the attractive blossoms. Demand was heightened because the bulb of a strikingly patterned flower might degenerate or rapidly die. Various methods of cultivation were alleged to increase breaking, but the capricious cause was unknown. A colossal gambling craze developed in Holland in 1634-1637 called tulipomania." Fortunes were traded for single prized bulbs. The episode was finally stopped by the government (Frylink, 1954). We now know that the virus is transmitted mechanically by several aphids, including Myzus persicae.

Viruses pathogenic to vertebrates that are transmitted by arthropods are collectively called "arboviruses," an abbreviation of "arthropodborne viruses." These contain RNA as the nucleic acid. Out of more than 200 arboviruses, 50 infect humans (Horsefall and Tamm, 1965). Yellow fever, dengue, and various encephalitis viruses are the most important. Mosquitoes are the vectors for the great majority of arboviruses, followed by ixodid ticks, Culicoides flies (Ceratopogonidae), Phlebotomus flies (Psychodidae) and possibly laclapid mites.

Yellow fever is an arbovirus transmitted by the mosquito Aedes aegypti. This disease may be fatal in up to 10 per cent of human cases. Hundreds of thousands have died, especially in tropical Africa and South America. The yellow fever mosquito was originally from Africa, but now occurs worldwide as a vector in urban environments. Although controlled in cities and towns, the virus persists in tropical forests. Here it is transmitted among monkeys by forest mosquitoes such as Haemagogus in the New World and various Aedes in Africa. Jungle yellow fever is transmitted to humans living in forest clearings and working in forests.

Insects in Relation to Bacteria, Spirochetes, and Rickettsiae

The classification of procaryotic organisms is continually revised as new information is obtained (Buchanan and Gibbons, 1974). Among procaryotes broadly called the bacteria are distinctive groups closely associated with arthropods: certain true or eubacteria, spirochetes, rickettsias, and mycoplasmas. The last group was mentioned in the preceding discussion of viruses because, until recently, insect-borne mycoplasmas have been confused with certain persistent plant viruses. The procaryotes in this section all have cell walls.

Eubacteria are commonly found on the integument, in the gut, and in mycetocytes. Endosymbiotic bacteria occur in Orthoptera, Isoptera, Phthiraptera, Hemiptera, Coleoptera, Hymenoptera, and Diptera. Certain larval calliphorid flies, known as "surgical maggots," are exceptional in that the gut is sterile. A bactericidal substance, allatonin, kills bacteria in the midgut. Napoleon's surgeon, Larrey, noted that healing was actually enhanced when neglected battle wounds were infected with maggots. Maggots were later used medically because they ingest bacteria and necrotic tissues.

Bacteria pathogenic to insects have been reviewed by Faust (1974). Infection is usually through ingestion. Important examples of bacterial diseases are milky disease of Japanese beetles, caused by Bacillus popilliae and B. lentimorbus, and the honeybee diseases, American foulboard, caused by Bacillus larvae, and European foulbrood, caused by Streptococcus pluton. Bacillus cereus and its variant, B. thuringiensis, produce crystals toxic to more than 182 species of pest insects, mostly Lepidoptera caterpillars. Commercial preparations of bacteria and crystals are applied with regular spray equipment to trees and crop plants in the control of pests.

Bacterial diseases of plants are mostly transmitted mechanically by insects, but some seem to be endosymbionts of the gut and may be transovarially transmitted. Fire blight of pears, apples, and some 90 kinds of other trees is caused by Erwinia amylovora and mechanically transmitted by many insects, especially bee pollinators.

Human diseases caused by bacteria and mechanically transmitted by insects are dysentery bacteria, Shigella, and Salmonella, carried by muscid filth flies, and tularemia, caused by Francisella tularensis and transmitted by the bite of tabanid deerflies.

The organism causing plague, or Black Death, Yersinia pestis, is transmitted biologically by fleas. Plague bacilli are naturally transmitted among wild rodents by rodent fleas. While hunting or camping human beings may acquire infections by handling diseased rodents or rabbits. Urban populations are endangered when the pathogen infects domestic rodents such as the black rat, Rattus rattus, and its ectoparasite, the oriental rat flea, Xenopsylla cheopis. Bacteria multiply in the flea's gut, forming a gelatinous plug in the proventriculus. A "blocked flea" is unable to pass blood to the midgut and repeatedly atttempts to suck. The result is that plague bacilli are regurgitated into the wounds. As the disease spreads among rats, the susceptible rats die. Fleas leave cold rats and transfer to live rats or humans. Following infection of a human by flea bites, the lymph nodes become inflamed—hence the name "bubonic plague," referring to the swollen buboes. In the most deadly phase of an epidemic, bacilli are rapidly spread from human to human by inhalation of infected respiratory droplets.

Plague has devastated the populations of entire countries since ancient times and has played a decisive role in history. In addition to the staggering statistics of human deaths, the impact of this disease is reflected in works of art and literature. In The Decameron, Giovanni Boccaccio (1313-1375) describes plague in Florence in 1348. This was at the beginning of the great epidemic that swept away 25 million Europeans and contributed to social change in the fourteenth century.

The terror of plague is shown in art beginning in 1348 (Brossollet, 1971). Reproduced here is an engraving. "The Pest," by Pierre Mignard (1610-1695). The scene was created from Biblical accounts of the plague of David (II Samuel 24 and I Chronicles 21) and Thucydides's description of the plague in Athens. Accurately illustrated are symptoms of unquenchable thirst (fountains in center background), insanity (delirious patient at left being restrained), and axillary buboes ruptured of lanced (armpit of woman in center foreground). The angel in the sky is pouring sulphurous fumes as a disinfectant. Smoke from torch and brazier were also intended to be purifying.

Spirochetes are slender, motile bacter.a of spiral configuration. Some are associated with insects. Termites of the advanced Family Termitidae may depend on spirochetes as intestinal endosymbionts. In the same genus as the spirochetes of syphilis, Treponema, are pathogens causing human skin diseases called pinta and yaws. These are transmitted mechanically from lesions by Hippelates (Chloropidae) eye gnats.

Rickettsiae are small, nonmotile bacteria that are spherical or rod-shaped. All are associated with arthropods at some point in their natural cycle. Diptera, Hymenoptera, and ticks possess commensalistic or mutualistic forms. The best-known rickettsiae are pathogens of vertebrates that are transmitted by bloodsucking fleas, lice, ticks, or mites. Epidemic, or louse-borne, typhus fever is caused by Rickettsia prowazeki and transmitted by human body lice, Pediculus humanus (Pediculidae). The microbes multiply in the louse's gut, passing out with feces. Human infection results when infected feces are inhaled or lice are crushed and the pathogens are introduced by fingers into skin abrasions or the eyes. Louseborne typhus spreads rapidly during times of social strife. Crowding and poor sanitation, as in war or famine, favour louse infestations. Typhus was partly responsible for the collapse of Napoleon's army in 1812.

Murine, or flea-borne, typhus is caused by Rickettsia typhi. The disease is transmitted among rats by rat fleas such as Xenopsylla cheopis (Pulicidae) and others. Humans acquire infection in the same manner as in louse-borne typhus, *i.e.*, through infected flea feces and crushed fleas. Other important rickettsial diseases are transmitted by acarines: Rocky Mountain spotted fever caused by Rickettsia rickettsii and transmitted by ticks; and scrub typhus, caused by Rickettsia tsutsugamushi and transmitted by chiggers.

Until recently rickettsias have been considered confined to insect or vertebrate hosts. Several diseases of plants, previously thought to be persistent viruses, are now associated with rickettsialike organisms: Pierce's disease of grapes, phony peach disease, and clover club leaf (Whitcomb, 1973). All are transmitted by leafhoppers. Clover club leaf has been shown to multiply in both the plant and insect.

Insects in Relation to Protozoa

Protozoa are endosymbionts in nearly all orders of insects but are best known in Orthoptera and Isoptera. Extraordinary quantities are found in termites, where flagellates constitute up to one-third of a nymph's weight. Protozoa enter insects via the month, and many types remain associated with the gut, but others become intracellular parasites or live in the hemocoel. They may be pathogenic, commensalistic, or mutualistic in relation to the insect host. Pathogens causing several major diseases of vertebrates are transmitted by insects, as are a few minor protozoan pathogens of plants.

The geologic record gives no clues to the antiquity of insect-symbiont relationships. That these associations can be quite old is inferred from the distribution of flagellates among the primitive termites and Cryptocercus roaches. Cleveland and his associates (1934) found 25 species of flagellates in Cryptocercus. Several species of the flagellate genus Trichonympha are found in both the roach and various termites. Because few opportunities for exchange of endosymbionts now seem to exist between the two orders of insects, we infer that both probably derive their Trichonympha from a common ancestor in the Mesozoic Era.

Protozoa pathogenic to insects have been reviewed by Brooks (1974). The most important are the microsporidians, which are parasites of many, if not most insects, especially of Lepidoptera and Diptera. The pebrine disease of silkworms, caused by Nosema bombycis, and nosema disease of honeybees, caused by N. apis, are examples. The coccidian Adelina is pathogenic in Coleoptera and Diptera. Relatively harmless are the eugregarines that are also known from many insects, especially Coleoptera, Orthoptera, and Diptera. Some are easily seen with the unaided eye because they may be up to 16 mm long. Mattesia grandis is, however, highly pathogenic to the boll weevil Anthonomis grandis, and may prove useful in its control.

At one time malaria was considered the most important disease of humans. It is caused by four species of sporozoans in the genus Plasmodium and transmitted by mosquitoes of the genus Anopheles. Persistent chemical insecticides, especially

DDT, and antimalarial drugs have substantially reduced mortality. The prospect was so encouraging that in 1955, the World Health Organization resolved to undertake a worldwide eradication programme. The use of residual insecticides continues to be the main strategy, but a variety of other methods are being developed to control resistant populations of the vectors. Unfortunately, malaria is now increasing in several parts of the world.

The biological transmission of plasmodia is of the cyclopropagative type. Female Anopheles ingest microgametocytes and macrogametocytes along with blood from infected humans. In the mosquito's midgut, the microgametocytes produce male microgametes that seek and penetrate the female cell, now enlarged and called a macrogamete. The resulting zygote develops into a motile ookinete. This enters the midgut epithelium and encysts as an oocyst between the epithelium and the outer tissues of the gut. Within the oocyst the cell undergoes mitotic and meiotic divisions, ultimately producing thousands of haploid sporozoites. The oocyst bursts, releasing the sporozoites in the hemocoel. The sporozoites invade various tissues, but those reaching the salivary glands are passed into another human's blood when the mosquito feeds. In the human, the sporozoites first enter cells of the liver, then produce merozoites that enter red blood cells. Merozoites develop into trophozoites that destroy the blood cell and release either more merozoites or gametocytes. The release of merozoites and their toxins occurs at regular periods according to the species of Plasmodium. This is the cause of the recurring fevers and chills that characterize malarial attacks.

Other important diseases of humans in tropical regions are caused by flagellates of the genera Trypanosoma and Leishmania. Several kinds of diseases are produced by the latter, including kala azar and espundia. Phlebotomus flies (Psychodidae) are known as vectors. Trypanosomes cause trypanosomiasis, or sleeping sickness. In Africa, both sexes of tsetse flies, Glossina (Muscidae), feed on blood and ingest flagellates. Some mechanical transmission may occur immediately by infected mouthparts, but cyclopropagative transmission follows. The flagellates multiply in the gut, and moving anteriorly to the salivary glands, they are passed to new hosts when the fly feeds. Trypanosomiasis in America is called Chagas' disease. The vectors are blood-sucking reduviid bugs of the Subfamily Triatominae. The transmission is also cyclopropagative in the insect's gut, but infection of new hosts is by feces. The bug defecates after feeding. Humans accidentally rub the contaminated feces into the wound or into the eyes or mouth.

Flagellates of the genus Phytomonas infest plants, especially those with latex. Vectors are various phytophagous Hemiptera-Heteroptera.

Insects in Relation to Fungi

Fungi have many interesting relationships with insects as commensals, mutuals, and pathogens. Many fungal diseases of plants are transmitted by insects, but apparently no pathogens of vertebrates.

The Laboulbeniales are fungi that live almost exclusively on the integument of Coleoptera, Diptera, and Neuroptera. As commensalistic ectosymbionts, they are transmitted from insect to insect by contact. The insect hosts are rarely killed by the fungi.

A mutualistic relationship exists between Septobasidium fungi and various scale insects in the Southeastern United States. Patches of fungus grow in concentric, annual rings on the bark of trees. Inside are tunnels and chambers enclosing scale insects. Thus embedded and protected, the scales feed on the plant with their long sucking mouthparts. The fungus apparently derives nourishment parasitically from the living insects and may kill the insect (Couch, 1938).

Mutualistic ectosymbionts among the fungi are frequently cultivated in "fungus gardens." Batra and Batra (1967), Francke-Grosmann (1967), Graham (1967), and Weber (1966) provide many details of the relationships. The insects are gall midges (Lasioptera, Cecidomyiidae), wood wasps (Siricidae and Xiphydriidae), ambrosia beetles (various Scolytidae, Platypodidae, and Lymexylidae), fungus ants (Atta, Formicidae), and fungus termites. Nutritions food is obtained from fungal saprophytes that grow on cellulose-rich wood or vegetable matter. Some fungus is carried by the adult insects to new locations. Queen Atta ants carry a pellet of fungus in a pouch in the mouth when they fly forth to establish new colonies. Wood-boring ambrosia beetles and wood wasps have special integumental cavities called mycetangia. Here the fungus is protected from desiccation and is nourished by a secretion. The mycetangia are filled with fungus before the insect leaves its infected gallery. When the beetles bore into a new host or the wasps oviposit, the fungus is inoculated in the area where the progeny will develop.

Blue-stain fungi of the genus Ceratocystis are associated with certain scolytid bark beetles. The spores are carried on the body, in mycetangia, or passed with feces. Sapwood moisture of the infected tree is reduced, making the tree more favourable for beetles. The stain lowers the commercial value of the wood.

Dutch elm disease is caused by Ceratocystis ulmi. First discovered in Ohio in 1930, this European disease has killed or has made necessary the destruction of elm trees in many urban areas. It has been recently recognized in California. Scolytus multistriatus and Hylurgopinus rufipes are vectors. The former feeds on twigs of healthy trees after emergence from infected trees or logs. The beetle then seeks sick trees or freshly cut elm wood to construct its breeding galleries. The healthy trees are infected with spores that cling to the insect. The fungus blocks water-conducting tissues, causing limbs or the whole tree to wilt. The afflicted parts then become suitable breeding sites for more beetles.

Endosymbiotic fungi and yeasts are found in the mycetomes of Hemiptera-Homoptera and intestines of anobiid and cerambycid beetles.

Fungi that are pathogenic to insects normally invade through the integument, but a few enter via the gut. Warm humid environments favour fungal attack. The fungal filaments and reproductive structures are often evident on the body surface. Species of Entomophthora and Beauveria are common pathogens. In the past, the latter caused serious losses of silkworms in China and Europe. Beauveria globulifera was early noted as causing a contagious disease of the chinch bug Blissus leucopterus (Lygaeidae). Near the end of the last century, the fungus was mass-produced and distributed to farmers for control of the pest. Entomophthora infects all life stages of insects, especially Diptera, Hemiptera, Lepidoptera, Coleoptera, and Orthoptera. In the fall, Entomophthora muscae causes epidemics among flies, leaving them attached in lifelike postures inside houses or on plants.

Insects mechanically transmit fungal pathogens of plants, but this does not preclude intimate biological relationships. The Dutch elm disease mentioned above is an example. Spores of Fusarium moniliforme, causing the disease endo-sepsis of figs, are transmitted when the fig wasp pollinates the flowers. A honeydew-like secretion is caused by the ergot fungus Claviceps, when it attacks grasses and cereals. The secretion, infected with spores, is rich in amino acids and attractive to Diptera. The insect visitors spread disease by carrying the spores externally or by passing them in feces.

Insects in Relation to Helminths

Wormlike animals are called helminths. We are concerned here with the phyla Platyhelminthes, Aschelminthes, and Acanthocephala. You may wish to renew your knowledge of these phyla by consulting a recent text in zoology (Storer, *et al.*, 1972) or animal parasitology (Olsen, 1974; Dunn, 1969).

Among the flatworms, or Platyhelminthes, are two classes that involve insects as intermediate hosts. The digenean flukes of the class Trematoda usually depend on snails as the first intermediate host, then enter a second intermediate host (snails, crustaceans, or fish) or encyst on aquatic plants before being ingested by the definitive vertebrate host.

The fluke Prosthogonimus spp., however, seeks drogonfly naiads as the second intermediate host after parasitizing snails. The parasites enter the anus while the naiad takes in water during respiration. Birds eat the infested naiads or adult dragonflies and become the definitive hosts.

One of the most remarkable eflects of a parasite on its vector has been observed in the case of the fluke Dicrocoelium spp. Eggs are released by the adult flukes in the definitive host and exereted. The eggs are then ingested by land snails and the parasites later expelled from the snails' respiratory pore in a ball of slime. Hundreds of parasites may be protected in each ball. The balls are collected by several species of ants in the genera Formica and Proformica. In the nest, the slime balls are eaten by the colony, resulting in a high rate of infection. Most parasites encyst in the ants' abdomen, bnt one or two often migrate to the brain, where a lesion develops. In the evening, ants usually descend from vegetation, but those with brain lesions remain on the vegetation until warmed by the next day's sun. Ruminants grazing early in the day ingest the infected ants, thus completing the cycle.

The tapeworms, or Cestodea, depend primarily on arthropods as intermediate hosts. Crustaceans are the usual hosts because most tapeworms are marine. The adults occupy the intestines of vertebrates. Some tapeworms in the terrestrial environment utilize insects as intermediate hosts. Eggs of Dipylidium caninum, the dog tapeworm, are ingested by larval dog fleas, Ctenocephalides canis (Pulicidae); cat fleas, C. felis; human fleas; Pulex irritans (Pulicidae); and the dog louse, Trichodectes canis Crichodectidae). The larval fleas and the biting lice have chewing mouthparts. They are able to ingest eggs, whereas the adult fleas cannot. Dogs and cats and their wild relatives acquire tapeworms by eating adult fleas or lice that were infected in earlier instars. Humans are occasionally hosts.

Various species of Hymenolepis infect aquatic crustaceans, earwigs, dung and scavenger beetles, millipeds, meal moths, and fleas. Hymenolepis nana, the dwarf tapeworm of humans, mice, and rats, and H. diminuta, another topeworm of mice and rats, both depend on insects as intermediate hosts. Coprophagic scavengers, including larval fleas and the mealworm beetle, Tenebrio molitor, ingest infected rodent feces. The insect is eaten in turn by the definitive vertebrate hosts. Humans may be parasitized also. A numer of other genera of tapeworms involve beetles, grasshoppers, larval and adult flies, and ants in their life cycles (Dunn, 1969).

The spiny-headed worms, or Acanthocephala, also live as endoparasitic adults in the intestines of terrestrial and aquatic vertebrates. On land, the eggs are ingested by insects, and in water probably by both insects and crustaceans. The intermediate hosts are later eaten by the vertebrate. Macracanthorhynchus hirudinaceus is a parasite of pigs and their wild relatives, and occasionally humans. The insect hosts are coprophagic beetle larvae of the Families Scarabaeidae, Tenebrionidae, and Hydrophilidae. They acquire eggs of the parasite be eating pig feces. The pigs then eat the infected larvae in the course of rooting in the soil.

The Aschelminthes include two classes of worms that infect insects: roundworm or Nematoda, and hairworms or Nematomorpha. The immature stages of the latter are endoparasitic in insects and crustaceans. Adult hairworms are free-living in saltwater or freshwater. The complete cycle in terrestrial and freshwater habitats is not well known. The adult worms copulate, lay eggs, and die in water. Larval worms emerge and may either penetrate soft-bodied aquatic invertebrates or be ingested by aquatic and possibly by terrestrial insects. Some hosts are not suitable for further development. In these the larvae encyst. Development proceeds, however, in larger terrestrial insects that acquire infection directly from water or by eating in intermediate hosts infested with encysted larvae. The parasite enters the hemocoel, feeds, and matures. When the host is near or in water, the adult worm emerges from the host quickly and becomes free-living. Heavily parasitized hosts may have the reproductive organs reduced in size. Emergence of the worm is fatal to the host.

May (1919) found up to 20 per cent of tettigoniid grasshoppers near Urbana, Illinois, infected with Gordius robustus. Paragordius varius was found in gryllid crickets. May believed the parasites were obtained directly from water. Inoue (1962) traced part of the life cycle of Chordodes in Japan. Mayfly naiads ingest the larvae, and these encyst. The infected adult mayflies are eaten in turn by larger predators such as the praying mantis. The cyst is digested, freeing the larva to penetrate the host's gut. Development is completed in the hemocoel.

The most numerous helminths associated with insects are nematodes. Much remains to be learned about the relationships between the largest group of animals, *i.e.*, the insects, and what is possibly the second largest group, the nematodes (Welch, 1965; Poinar, 1975). Nematodes have essentially five kinds of relationships to insects:

(1) phoresy,

(2) facultative parasitism,

(3) obligatory parasitism that depends only on insects as hosts,

(4) parasitism involving both insects and plants as hosts, and

(5) obligatory parasitism with insects as intermediate hosts and vertebrates as definitive hosts.

Insects are vectors of important nematode diseases of humans and other animals, but they are curiously not vectors of nematodes causing plant diseases.

Insects regularly serve to transport nematodes in a phoretic relationship. Juvenile stages of rhabditoid nemateodes in decaying matter or beetle galleries in wood attach themselves externally to insect scavengers or borers and are carried to new habitats. Some harmlessly enter the insect's body, effectively avoiding desiccation.

The internal phoretic relationship becomes one of facultative parasitism when the nematode feeds on the host's tissues, out is also able to complete its life cycle without an insect host (Poinar, 1972). The effect on the host varies from no apparent harm to death. Species of Neoaplectana have an odd relationship to their insect hosts that defies classification. The juvenile worm penetrates the insect's body like a true endoparasite. Specific bacteria are released by the worm after it enters the hemocoel. The bacteria rapidly kill the host, and the nematode completes up to several generations, feeding on the dead host's tissues and bacteria. Neoaplectana is probably obligatorily associated with insects in nature, yet can be reared on artificial media such as dog food. Is it a facultative parasite, predator, or saprophyte? Regardless of label, the nematodes are among the most promising organisms for use in the control of insect pests.

Obligate parasites feed only on insects. The juvenide stages of obligate nematode parasites of insects develop in the host's hemocoel, intestine, or reproductive organs. Adults may be free-living or remain in the host. One important family are the mermithids that infest aquatic insects and terrestrial insects in moist habitats. The host is killed when the nematode emerges. Infection is sufficiently high in some areas to reduce significantly or to eradicate the host populations. Nematodes influence the host's development by destroying organs, removing nutritional reserves, or possibly producing toxic effects on the corpora allata. Altered behaviour, reduced fecundity, sterilization, inrersexes, and intercastes of social insects may result from parasitism (Welch, 1965).

Juvenile nematodes of the genus Deladenus infest the hemocoel of siricid wood wasps. The adult parasites mature outside the host. Two types of females are produced. Ona type does not feed and after mating, penetrates a new host.

The other type feeds on a fungal symbiote of the wood wasp. Female wood wasps inoculate the fungus where they oviposit. Both sexes of the wood wasps are sterilized by the parasites, but the female continues to fly to new trees for oviposition. The fungus-feeding nematodes are deposited during the futile oviposition in new places where other wood wasps are present (see Poinar, 1972). A related nemarode is parasitic on weevils and mustard plants.

The last group of nematodes involve species whose definitive hosts are vertebrates. These are mainly the filarial worms, some of which are host specific to humans. The adult worms feed on lymph and tissue fluids. The young or microfilariae are produced viviparously. Blood-sucking Diptera ingest the microfilariae. These penetrate the insect's gut and become intracellular parasites in fat, muscle, or Malpighian tubule cells. In 1 to 2 weeks, they metamorphose into an infective stage that migrates to the proboscis. When the insect feeds, the parasites migrate into the wound.

The grossly swollen legs, breasts, or genitalia characteristic of elephantiasis are caused by Wuchereria bancrofti. Culex pipiens quinquefasciatus (Culicidae) is the principal vector, but various species in other mosquito genera are also found infected Species of Simulium (Simuliidae) transmit Onchocerca volvulus, a blinding disease of the Tropics. Another eye disease in Africa, caused by Loa loa, is carried by deerflies of the genus Chrysops (Tabanidae).

Other nematodes combining insects and vertebrates as hosts include the ascaridoid Subulura spp. that infects beetles and cockroaches, and then galliform birds. Many genera of spiruroid nematodes utilize coprophagous scarab beetles, cockroaches, termites, and anthomyiid flies as intermediate hosts. Definitive hosts are domestic animals and birds and their wild allies (Dunn, 1969).

11

Insects and Man

Introduction

In this final chapter we shall focus on those insects that Man describes, in his economically minded way, as beneficial or harmful, though it should be appreciated from the outset that these constitute only a very small fraction of the total number of species. Further, it must also be realized that the ecological principles that govern the interactions between insects and Man are no different from those between insects and any other living species, even though Man with his modern technology can modify considerably the nature of these interactions.

Of an estimated 3 million species of insects, probably not more than about 15,000 (0.5%) interact, directly or indirectly, with Man. Perhaps some 3000 species constitute pests that, either alone or in conjunction with microorganisms, cause significant damage or death to Man, agricultural products, and manufactured goods (Williams, 1967). In the United States alone, for example, it is estimated that harmful insects annually do about 5 billion dollars' worth of damage (Borror *et al.*, 1976). On the other hand, according to Borror *et al.*, the value of benefits derived from insects each year amounts to more than 7 billion dollars. That insects do more good than harm probably would come as a surprise to lay persons whose familiarity with insects is normally limited to mosquitoes, houseflies, cockroaches, various garden pests, etc., and to farmers who must protect their livestock and crops against a variety of pests. If asked to prepare a list of useful insects, many people most likely would not get farther than the honeybee and, perhaps, the silkmoth, and would entirely overlook the enormous number of species that act as pollinating agents or prey on harmful insects which might otherwise reach pest proportions.

Man has long recognized the importance of insects in his well-being. Insects and/or their products have been eaten by Man for thousands of years. Production of silk from silk worm pupae has been carried out for almost 5000 years. Locust swarms which originally may have been an important seasonal food for Man, took on new significance as he became a farmer rather than a hunter. However, except for the honeybee and silk worm, whose management is relatively simple and labour-intensive, until recently Man neither desired nor was able, due to his lack of basic knowledge as well as of technology, to attempt large-scale modification of the environment of insects, either to increase the number of beneficial insects or to decrease the number of those designated as pests.

Several features of Man's recent evolution have made such attempts imperative. These include a massive increase in human population, a trend toward urbanization. increased geographic movement of people and agricultural products, and, associated with the need to feed more people, a trend toward monoculture as an agricultural practice.

The relatively crowded conditions of urban areas enable insects parasitic on humans both to locate a host (frequently a prerequisite to reproduction) and to transfer between host individuals. Thus, urbanization facilitates the spread of insect-borne human diseases such as typhus, plague, and malaria whose spectacular effects on human population are well documented. For example, in the sixth century A.D. plague was responsible for the death of about 50 per cent of the population in the Roman Empire, and "Black Death" killed a similar proportion of England's population in the mid-1300s (Southwood, 1977).

An increasing need to produce more and cheaper food led, through agricultural mechanization, to the practice of monoculture, the growing of a crop over the same large area of land for many years consecutively. However, two faults of monoculture are (1) the ecosystem is simplified, and (2) as the crop plant is frequently graminaceous (a member of the grass family, including wheat, barley, oats, rice, and corn), the ecosystem is artificially maintained at an early stage of ecological succession. By simplifying the ecosystem, Man encourages the buildup of populations of the insects that compete with him for the food being grown. Further, as the competing insects are primary consumers, that is, near the start of the food chain, they typically have a high reproductive rate and short generation time. In other words, populations of such species have the potential to increase at a rapid rate.

A massive increase in Man's geographic movements and a concomitant increase in trade, led to the transplantation of a number of species, both plant and animal, into areas previously unoccupied by them. Some of these were able to establish themselves and, in the absence of normal regulators of population (especially predators), increased rapidly in number and became important pests. Sometimes, as Man colonized new areas, some of the cultivated plants he introduced proved to be an excellent food for species of insects endemic to these areas. For example, the Colorado beetle, Leptinotarsa decemlineata, was originally restricted to the southern Rocky Mountains and fed on wild Solanaceae. With the introduction of the potato by settlers, the beetle had an alternate, more easily accessible source of food, as a result of which both the abundance and distribution of the beetle increased and the species became an important pest. Likewise, the apple maggot, Rhagoletis pomonella, apparently fed on hawthorn until apples were introduced into the eastern United States (Horn, 1976).

2. Beneficial Insects

Insects may benefit Man in various ways, both directly and indirectly. The most obvious of the beneficial species are those whose products are commercially valuable. Considerably more important, however, are the insects that pollinate crop plants. Other beneficial insects are those which are used as food, in biological control of pest insects and plants, and in research. For some of these useful species, Man modifies their environment in order to increase their distribution and abundance so that he may gain the benefits.

Insects whose Products are Commercially Valuable

The best-known insects that fall in this category are the honeybee (Apis mellifera), silkworm (Bombyx mori), and lac insect (Laccifer lacca).

The honeybee originally occupied the African continent, most of Europe (except the northern part), and western Asia, and within this area the usefulness of its products, honey and beeswax, has been known for many thousands of years. Though the discovery of sugar in cane (in India, about 500 B.C.) and in beet (in Europe, about 1800 A.D.) (Southwood, 1977) led to a decline in the importance of honey, it is nevertheless still a very valuable product.

Bee management was probably first practiced by the ancient Egyptians. Honeybees were brought to North America by colonists in the early 1600s, and today honey and beeswax production is a multimillion dollar industry. Good bee management aims to maintain a honeybee colony under optimum conditions for maximum honey and wax production. Management details vary according to the climate and customs of different geographical areas but may include (1) moving hives to locations where nectar-producing plants are plentiful; (2) artificial feeding of newly established, spring colonies with sugar syrup in order to build up colony size in time for the summer nectar flow; (3) checking that the queen is laying well and, if not, replacing her; (4) checking and treating colonies for diseases such as foulbrood and nosema; and (5) increasing the size of a hive as the colony develops, in order to prevent swarming.

Silk production has been commercially important for about 4700 years. The industry originated in East Asia and spread into Europe (France, Italy, and Spain) after eggs were smuggled from China to Italy in the sixth century A.D. The development of cheap synthetic fibres has had a severely deleterious effect on the silk industry which is labour-intensive, making production costs high. Nevertheless, the annual world production is about 30-35 million kilograms (Borror *et al.*, 1976).

The lac insect is a scale insect endemic to India and southeast Asia that secretes about itself a coating of lac, which may be more than 1 cm thick. The twigs on which the insects rest are collected and either used to spread the insects to new areas or ground up and heated in order to separate the lac. The lac is a component of shellac, though its importance has declined considerably with the development of synthetic materials.

2.2. *Insects as Pollinators*

An intimate, mutualistic relationship has evolved between many species of insects and plants, in which plants produce nectar and pollen for use by insects, while the latter provide a transport system to ensure effective cross-pollination. Though some crop plants are wind-pollinated, for example, graminaceous species, a large number, including fruits, vegetables, and field crops such as clovers, rape, and sunflower, require the service of insects. In addition, ornamental flowers are almost all insect pollinated.

The best-known, though by no means the only, important insect pollinator is the honeybee, and it is standard practice in parts of the world for fruit and vegetable growers either to set up their own beehives in their orchards and fields or to contract this job out to beekeepers in the vicinity. Indeed, it is apparent that under such conditions the value of the bee as a pollinator may be 15 to 20 times its value as a honey producer. Certainly, the presence of bees in orchards, alfalfa fields, etc., may result in a three or fourfold increase in the yield of fruit or seed. Borror *et al.* (1976) estimate the annual value of insect-pollinated crops to be about 6 billion dollars.

2.3. Insects as Agents of Biological Control

It is only relatively recently that Man has gained an appreciation of the importance of insects in the regulation of populations of potentially harmful insects and plants. In many instances, this appreciation was gained only when, as a result of Man's own activities, the natural regulators were absent, a situation which was rapidly exploited by these organisms whose status was soon elevated to that of pest. In the first three examples given below (taken from DeBach, 1974), none of the organisms is a pest in its country of origin due to the occurrence there of various insect regulators. The discovery of these regulators, followed by their successful culture and release in the area where the pest occurs, constitutes biological control.

Probably the best-known examples of an introduced plant pest are the prickly pear cacti (Opuntia spp.) taken into Australia as ornamental plants by early settlers. Once established, the plants spread rapidly so that by 1925 some 60 million acres of land were infested, mostly in Queensland and New South Wales. Surveys in both North and South America, where Opuntia spp. are endemic, revealed about 150 species of cactus-eating insects, of which about 50 were judged to have biological control potential and were subsequently sent to Australia for culture and trials. Larvae of one species, Cactoblastis cactorum, a moth, brought in from Argentina in January 1925, proved to have the required qualities and within 10 years had virtually destroyed the cacti. Perhaps the most remarkable feature of this success story is that only 2750 Cactoblastis larvae were brought to Australia, of which only 1070 became adults. From these, however, more than 100,000 eggs were produced, and in February—March of 1926 more than 2.2 million eggs were released in the field! Additional releases, and redistribution of almost 400 million field-produced eggs until the end of 1929, ensured the project's success.

The classical example of an insect pest brought under biological control is the cottony-cushion scale. Icerya purchasi, which was introduced into California, probably from Australia, in the 1860s. Within 20 years, the scale had virtually destroyed the recently established, citrus-fruit industry in Southern California. As a result of correspondence between American and Australian entomologists and of a visit to Australia by an American entomologist, Albert Koebele, two insect species were introduced into the United States as biological control agents for the scale. The first, in 1887, was Cryptochaetum iceryae, a parasitic fly, about which little is heard, though DeBach (1974) considers that it had excellent potential for control of the scale had it alone been imported. However, the abilities of this species appear to have been largely ignored with the discovery by Koebele of the vedalia beetle, Vedalia cardinalis, feeding on the scale. In total, only 514 vedalia were brought into th United States, between November 1888 and March 1889, to be cultured on caged trees infested with scale. By the end of July 1889, the vedalia had reproduced to such an extent that one orchardist, on whose trees about 150 of the imported beetles had been placed for culture, reported having distributed 63,000 of their descendents since June 1! By 1890, the scale was virtually wiped out. Similar successes in controlling scale by means of vedalia or Cryptochaetum have been reported from more than 50 countries (DeBach, 1974).

As a third example of an introduced pest being brought under control by biological agents, we may cite the winter moth, Operophtera brumata, which, though endemic to Europe and parts of Asia, was accidentally introduced into Nova Scotia in the 1930s. Its initial colonization was slow, and it did not reach economically significant proportions until the early 1950s, and by 1962 it had spread to Prince Edward Island and New

Brunswick. The larvae of the winter moth feed on the foliage of hardwoods such as oak and apple. Though more than 60 parasites of the winter moth are known in western Europe, only six of these were considered to be potential control agents and introduced into esstern Canada between 1955 and 1960. Only two of these, Cyzenis albicans, a tachinid, and Agrypon flaveolatum, an ichneumonid, became established, but between them they brought the moth under control by 1963. Embree (in Huffaker, 1971) rotes that the two parasites are both compatible and supplementary to each other. When the density of moth larvae is high, C. albicans which is attracted to feeding damage caused by the larvae, is a more efficient parasite than A. flaveolatum. However, once in the vicinity of damage, it does not specifically seek out winter moth larvae. Thus, at lower density, it wastes eggs on nonsusceptible defoliators such as caterpillars of the fall cankerworm. Alsophila pometaria. Hence, at low host densities, A. flaveolatum is more effective because it oviposits specifically on winter moth larvae.

The three, examples described above indicate one method whereby the importance of biological control can be demonstrated, namely, by introduction of potential pests into areas where natural regulators are absent. Another way of demonstrating the same phenomenon is to destroy the natural regulators in the original habitat, which enables potential pests to undergo a population explosion. This has been achieved frequently through the use of nonselective insecticides such as DDT. For example, the use of DDT against the codling moth, Carpocapsa pomonella, in the walnut orchards of California, led to outbreaks of native frosted scale, Lecanium pruinosum, which was unaffected by DDT, whereas its main predator, an encyrtid, Metaphycus californicus, suffered high mortality (Hagen *et al.*, in Huffaker, 1971). Another Lecanium scale. L. coryli, introduced from Europe in the 1600s, is a potentially serious pest of apple orchards in Nova Scotia but is normally regulated by various natural parasitoids (especially the chalcidoids Blastothrix sericea and Coccophagus sp.) and predators (especially mirid bugs). Experimentally it was clearly demonstrated in the 1960s that application of DDT destroyed a large proportion of the Blastothrix and mirid population, and this was followed in the next 2 years by medium to heavy scale infestations. Recovery of the parasite and predators was rapid, however, and by the third year after spraying the scale population density had been reduced to its original value (MacPhee and MacLellan, in Huffaker, 1971).

2.4. Insects as Food

In addition to the many predaceous insects which feed on other insects, many vertebrates are insectivorous to a variable degree. Some of these vertebrates, especially some freshwater fish and game birds, in turn, may be eaten by humans. In some areas of the world, insects "in season" routinely form part of the human diet. Locusts, termites, caterpillars, and grubs of various kinds are eaten in Africa; edible caterpillars are sold fresh or in cans in Mexico; and Australian aborigines are reported to eat roasted bugong moths (Agrotis infusa). Most North Americans and Europeans have not yet been educated to the delights (and nutritional value) of insects, and despite the efforts of authors such as Taylor and Carter (1976) to increase the popularity of insects as food, at present cans of edible insects are restricted to the gourmet counter of food stores.

2.5. Soil-Dwelling and Scavenging Insects

By their very habit the majority of soil-dwelling insects are ignored by Man. Only those which adversely affect his well-being, for example, termites, wireworms, and cutworms, normally "merit" his attention. When placed in perspective, however, it seems probable that the damage done by such pests is greatly outweighed by the benefits which soil-dwelling insects as a group confer. The benefits include aeration, drainage, and turnover of soil as a result of burrowing activity. Many species carry animal and plant material underground for nesting, feeding, and/or reproduction, which Metcalf *et al.* (1962) compare to ploughing in a cover crop.

Many insects, including a large number of soil-dwelling species, are scavengers; that is, they feed on decaying animal or plant tissues, including dung, and thus accelerate the return of elements to food chains. In addition, through their activity they may prevent use of the decaying material by other, pest insects, for example, flies. Perhaps of special interest are the dung beetles (Scarabaeidae), most species of which bury pieces of fresh dung for use as egg-laying sites. Generally, the beetles are sufficiently abundant that a pat of fresh dung may completely disappear within a few hours, thus reducing the number of dung-breeding flies that can locate it. Furthermore, the chances of fly eggs or larvae surviving within the dung are very low because the dung is ground into a fine paste as the beetles or their larvae feed. Likewise, the survival of the eggs of tapeworms, roundworms, etc., present in the dung producer, is severely reduced by this activity.

In Australia, which has a large population of cattle (derived largely from animals taken in by early settlers), cow dung presents a serious problem because, until recently, the beetle species that normally dispose of this kind of dung were absent. (Australia does have dung beetles, but these are specialized for dealing with the dung of kangaroos and other marsupials.) In addition to providing food for a massive number of fly larvae when freshly deposited, the dung, because of the dry climate, soon dries and then may remain unchanged for a considerable time. Rank herbage grows around each pad of dung but this is not normally eaten by cattle. Thus, at any time, a considerable proportion of all grazing land (estimated at 20 per cent or 6 million acres total) is not usable. In 1963, it was decided to initiate a programme of biological control of dung using various South African species of dung beetles. After extensive research in 1967 the first beetles were released in northern Australia. Several species have become established and are spreading rapidly down the eastern side of Australia. In general, the beetles are doing an excellent job of "cleaning up," except for the cooler months when they are inactive. Additional species, which are active at lower temperatures are being tested so as to obtain year round dung removal (Waterhouse, 1974).

2.6. Other Benefits of Insects

Their relatively simple food and other requirements, short generation time, and high fecundity enable many insects to be reared cheaply and easily under laboratory conditions and consequently, make them valuable in teaching and research. The fruit fly. Drosophila melonagaster, with its array of mutants, is familiar to all who take an elementary genetics class, though it must also be appreciated that the insect continues to have an important role in advanced genetic research. Studies on other insects have provided us with much of our basic knowledge of animal and cell physiology, particularly in the areas of nutrition, metabolism, endocrinology, and neuromuscular physiology. Grain beetles are good subjects for the study of population ecology.

Many insects give us pleasure through their aesthetic value. Because of their beauty, certain groups, especially butterflies, moths, and beetles, are sometimes collected as a hobby. Some are embedded in clear materials from which jewelry, paperweights, bookends, place mats, etc., are made. Others are simply used as models on which paintings and jewelry are based.

3. Pest Insects

Since Man evolved, insects have fed on him, competed with him for food and other resources, and acted as vectors of microorganisms that cause diseases in Man or the organisms of value to him. However, as was noted in the Introduction, the impact of such insects increased considerably as the human population grew and became more urbanized. Urbanization presented easy opportunities for the dissemination of Man's insect parasites and·the diseases they carry. Large-scale and long-term cultivation of the same crop over an area facilitated rapid population increases in certain plant-feeding species and the spread of plant diseases. Modern transportation, too, encourages the spread of pest insects and insect-borne diseases. Further, as described in Section 2.3. some of Man's attempts at pest eradication have backfired, resulting in even greater economic damage.

3.1 Insects Affecting Man Directly

A large number of insect species may be external, or temporary internal, parasites of Man. Some of these are specific to Man, for example, the body louse (Pediculus humanus) and public louse (Phthirus pubis), but most have a variable number of alternate hosts which compounds the problem of their eradication. With rare exceptions, for example, some myiasis-causing flies, insect parasites are not fatal to humans. In large numbers, insect parasites may generally weaken their host, making him more susceptible to the attacks of disease-causing organisms. Or the parasites, as a result of feeding, may cause irritation or sores which may then become infected.

But by far the greatest importance of insect parasites of Man is their role as vectors of pathogenic microorganisms (including various "worms") some well-known examples of which are given in Table 11.1. The pathogen is picked up when a parasitic insect feeds and may or may not go through specific stages of its life cycle in the insect. Bacteria and viruses are directly transmitted to new hosts, an insect serving as a mechanical vector, whereas for protozoa and worms (tapeworms and nematodes), an insect serves as an intermediate host in which an essential part of the parasites' life cycle occurs. In the latter arrangement the insect is known as a biological vector.

A pathogen may reside (and multiply) in alternate vertebrate hosts which are immune to or only mildly infected by it. For example, the bacterium Pasteurella pestis, which causes bubonic plague (Black Death), is endemic in wild rodent populations. However, in domestic rats and humans, to which it is transmitted by certain fleas, it is highly pathogenic. Similarly, in South America, yellow fever virus, transmitted by mosquitoes, is found in monkeys though these are immune to it. Such alternate hosts are thus an important reservoir of disease.

TABLE 11.1

Examples of Insects Which Serve as Vectors for Diseases of Man and Domestic Animals[1]

Insect vector	*Pathogen*	*Disease*	*Host*	*Distribution*
1	*2*	*3*	*4*	*5*
Anoplura				
Pediculus humanus (body louse)	Rickettsia prowazekii (rickettsian)	Epidemic typhus (Brills disease)	Man, rodents	Worldwide
	Pasteurella tularensis (bacterium)	Tularemia	Man, rodents	N. America, Europe, the Orient
Hemiptera				
Rhodnius spp. (assassin bugs)	Trypanosoma cruzi (protozoan)	Chagas' disease	Man, rodents	S. America, Central America, Mexico, Texas
Diptera				
Phlebotomus spp. (sand flies)	Leishmania donavani (protozoan)	Kala-azar (Dumdum fever)	Man	Mediterranean region, Asia, S. America
	L. tropica	Oriental sore	Man	Africa. Asia S. America

	L. braziliensis	Espundia	Man	S. America, Central America, N. Africa, southern Asia
	(virus)	Pappataci fever (sand fly fever)	Man	Mediterranean region, India, Sri Lanka
Anopheles spp. (mosquitoes)	Plasmodium vivax (protozoan)	Malaria	Man	Worldwide in tropical, subtropical, and
	P. malariae	Malaria	Man	temperate regions
	P. falciparum	Malaria	Man	
Aedes spp. (mosquitoes)	(virus)	Yellow fever	Man, monkeys, rodents	American and African tropics and subtropics
	(virus)	Dengue	Man	Worldwide in tropics and subtropics
	(virus)	Encephalitis	Man, horses	N. America, S. America, Europe, Asia
	Wucheria bancrofti (nematode)	Filariasis (Elephantiasis)	Man	Worldwide in tropics and subtropics
Culex spp. (mosquitoes)	(virus)	Dengue	Man	Worldwide in tropics and subtropics

(Contd.)

TABLE 11.1 (*Contd.*)

1	2	3	4	5
	(virus)	Encephalitis	Man, horses	N. America, S. America, Europe, Asia
	Wucheria bancrofti (nematode)	Filariasis (Elephantiasis)	Man	Worldwide in tropics and subtropics
Tabanus spp. (horse flies)	Bacillus anthroacis (bacterium)	Anthrax	Man, other animals	Worldwide
Chrysops spp. (deer flies)	Pasteurella tularensis (bacterium)	Tularemia	Man, rodents	N. America, Europe, the Orient
	Loa loa (nematode)	Loiasis (Calabar swelling)	Man	Africa
Glossina spp. (tsetse flies)	Trypanosoma rhodesiense (protozoan)	Sleeping sickness	Man, other animals	Equatorial Africa
	T. gambiense	Sleeping sickness	Man, other animals	Equatorial Africa
	T. brucei	Nagana	Cattle, wild ungulates	Equatorial Africa

Siphonaptera				
Xenopsylla cheopsis (oriental rat flea)	Pasteurella pestis (bacterium)	Bubonic plague (Black death)	Man, rodents	Worldwide
Xenopsylla spp.	Rickettsia typhi (rickettsian)	Endemic (murine) Typhus	Man, rodents	Worldwide
Xenopsylla cheopsis	Hymenolepis nana (cestode)	Tapeworm	Man	Europe and N. America
	H. diminuta (cestode)	Tapeworm	Man	Worldwide
Nosopsyllus fasciatus (northern rat flea)	Pasteurella pestis (bacterium)	Bubonic plague (Black Death)	Man, rodents	Worldwide
	Rickettsia typhi (rickettsian)	Endemic (murine) typhus	Man, rodents	Worldwide
	Hymenolepis diminuta (cestode)	Tapeworm	Man	Worldwide
Ctenocephalides canis (dogs, flea)	Dipylidium caninum (cestode)	Tapeworm	Man, dogs, cats	Worldwide
	Hymenolepis nana (cestode)	Tapeworm	Man	Europe, N. America
Pulex irritans (human flea)	Hymenolepis nana (cestode)	Tapeworm	Man	Europe, N. America

1. Data from Askaw (1971) and Horror *et al.* (1970).

Transmission of human disease-causing microorganisms is not, however, entirely the domain of parasitic insects. Many insects, especially flies, may act as mechanical vectors, contaminating human food as they rest or defecate on it, with pathogens picked up during contact with feces or other organic waste. Examples of such insects and the disease transmitted by them are listed in Table 11.1.

A third category of insects that directly affect Man includes those which may bite or sting when accidental contact is made with them, for example, bees, wasps, ants, some caterpillars (with poisonous hairs on their dorsal surface), and blister beetles. Normally, the effect of the bite or sting is temporary and nothing more than skin irritation, swelling, or blister formation. Bee stings, however, may cause anaphylaxis or death in some sensitive individuals.

3.2 Pests of Domesticated Animals

A range of insect parasites may cause economically important levels of damage to domestic animals. The majority of these parasites are external and include bloodsucking flies (for example, mosquitoes, horseflies, deerflies blackflies, and stable flies), biting and sucking lice, and fleas. Other parasites are internal for part of their life history, for example, bot-, warble, and screwworm flies, which as larvae live in the gut (horse bot), under the skin (warble and screwworm of cattle), or in head sinuses (sheep bot). Other examples are given in the chapters that deal with the orders of insects. In addition to insects, other arthropods are also important livestock pests, for example, various mites and ticks. For further details of the pests which attack particular domesticated animals, readers may consult Metcalf *et al.* (1962) and Pfadt (1971).

Generally the effect of such parasites is to cause a reduction in the health of the infected animal. In turn, this results in a loss of quality and/or quantity of meat, wool, hide, milk, etc., produced. When severely infected by parasites, an animal may eventually die. In addition to their own direct effect on the host, some parasites are vectors of livestock diseases, examples of which are included in Table 11.1.

3.3 Pests of Cultivated Plants

Damage to crops and other cultivated plants by insects is enormous and is valued at about 3 billion dollars annually in the United States (Borror *et al.*, 1976). Damage is caused either directly by insects as they feed (by chewing or sucking) or oviposit, or by viral, bacterial, or fungal diseases, for which insects serve as vectors. Especially important as "direct damagers" of plants are Orthoptera, Lepidoptera, Coleoptera, and Hemiptera, and the reader should consult the chapters that deal with these orders for specific examples of such pests. Several hundred diseases of plants are known to be transmitted by insects (for examples see Table 11.2) including about 300 which are caused by viruses (Eastop, 1977). Especially important in disease transmission are Hemiptera, particularly leafhoppers and aphids. Three aspects of the behaviour of these insects faciliate their role as disease vectors:

1. they make brief but frequent probes with their mouthparts into host plants;
2. as the population density reaches a critical level, winged migratory individuals are produced; and
3. in many species, winged females deposit a few progeny on each of many plants, from which new colonies develop.

On the basis of their method of transmission and viability (persistence in the vector), viruses may be arranged in three categories. The nonpersistent (style-borne) viruses are those believed to be transmitted as contaminants of the mouthparts. Such viruses remain infective in a vector for only a very short time, usually an hour or less. Semipersistent viruses are carried in the anterior regions of the gut of a vector, where they may multiply to a certain extent. Vectors do not normally remain infective after a molt, presumably because the viruses are lost

TABLE 11.2

Examples of Plant Diseases Transmitted by Insects[a]

	Disease	*Important hosts*	*Vectors*	*Distribution*
Viruses	Alfalfa mosaic	Alfalfa, tobacco, potato, beans, peas, celery, zinnia, petunia	Aphids (at least 16 spp.) Incl. Acyrthosiphon primulae, A. solani, Aphis craccivora, A. fabae. A. gossypii, Macrosiphum euphorbiae. M. pisi, Myzus ornatus, M. persicae, M. violae	Worldwide
	Barley yellow dwarf	Barley, oat, wheat, rye, wild and tame grasses	Numerous aphids incl. Macrosiphum granarium, M. miscanthi, Myzus circumflexus, Rhopalosiphum padi, R. maidis	North America, Australia, Denmark, Holland, U.K.
	Bean common mosaic	Beans	Aphids (at least 11 spp.) esp. Aphis rumicis, Macrosiphum pisi, M. gei	Worldwide

Beet yellows	Sugarbeet, spinach	Aphids esp. Aphis fabae, Myzus persicae	Wherever sugarbeet is grown
Cauliflower mosaic,	Cauliflower, cabbage, Chinese cabbage	Ahids esp. Brevicoryne brassicae, Rhopalosiphum pseudobrassicae, Myzus persicae	Europe. U.S.A., New Zealand
Dahlia mosaic	Dahlia, zinnia, calendula	Aphids esp. Myzus persicae, Aphis fabae, A. gossypii, Macrosiphum gei, Myzus convolvuli	Wherever dahlias are grown
Lettuce mosaic	Lettuce, sweet pea, garden pea, endive, aster, zinnia	Aphids esp. Myzus persicae, Aphis gossypii, Macrosiphum euphorbiae	Europe, U.S.A. (esp. California), New Zealand
Pea mosaic	Garden pea, sweet pea, broadbean, lupin clovers	Aphids: Acythosiphon pisi, Myzus persicae, Aphis fabae, A. rumicis	Europe, U.S.A., New Zealand, Austrialia, Japan,
Potato virus Y	Potato, tobacco, tomato, petunia, dahlia	Aphids esp. Myzus persicae, M. certus, M. ornatus, Macrosiphum euphorbiae	U.K., France. U.S.A.
Soybean mosaic	Soyboan	Aphids incl. Myzus persicae, Macrosiphum pisi	Wherever soybean is grown

(*Contd.*)

TABLE 11.2 (*Contd.*)

	Disease	*Important hosts*	*Vectors*	*Distribution*
	Sugarcane mosaic	Sugarcane, corn, sorgham. other tame and wild grasses	Numerous aphids incl. Rhopalosiphum maidis, Aphis gossypii, Schizaphis graminum, Myzus persicae	Wherever sugarcane is grown
	Tomato spotted wilt	Tomato, tobacco, dahlia, pineapple	Thrips: Thrips tobaci, Frankliniella schultzeri, F. fusca, F. occidentalis	Africa, Asia, Australia, Europe, North and South America
	Turnip yellow mosaic	Turnip, cauliflower, Chinese cobbage, kohlrabi, cabbage, broccoli	Flea beetles (Phyllotreta spp.): Mustard beetle (Phaedon cochleariae); Grasshoppers (Leptophyes punctatissima, Chorthippus bicolor); Earwig (Forficula auricularia)	U.K., Germany, Portugal, North America
Mycoplasmas	Aster yellows	Aster, celery, carrot, squash, cucumber, wheat, barley	Numerous leafhoppers incl. Gyponana hasto, Scaphytopius acutus, S. irroratus, Macrosteles fascifrons, Paraphlepsius apertnus, Texananus (several species)	Worldwide

	Clover phyllody	Most clovers	Leafhoppers incl. Aphrodes albifrons, Marcrosteles cristata, M. fascifrons, M. viridigriseus, Euscelis lineolatus, E. plebeja	U.S.A., U.K.
	Corn stunt	Corn	Leafhoppers esp. Dalbulus elimatus, D. maidis, Graminella nigrifrons	North and Central America
Bacteria	Stewarts bacterial wilt	Corn	Corn flea beetle (Chaetoceneme pulicara): toothed flea beetle (C. denticulata)	U.S.A.
	Cucurbit wilt	Cucumber, muskmelon	Cucumber beetles (Diabrotica vittata, D. duodecimpunctata)	U.S.A., Europe, South Africa, Japan,
	Potato blackleg	Potato	Seedcorn maggot (Hylemya cilicrura), H. trichodactyla	North America
	Fire blight	90 spp. of orchard trees and ornamentals, esp. apple, pear, quince	Wide range of insect vectors, esp. bees, wasps, flies, ants, aphids	North America, Europe
Fungi	Dutch elm disease	Elm	Elm bark beetles, esp. Scolytus multistratus, S. scolytus, Hylurgopinus rufipes	Asia, Europe, North America
	Ergot	Cereals and other grasses	About 40 spp. of insects esp. flies, beetles, aphids	Worldwide

[a]Data from various sources.

when the foregut intima is shed. Persistent (circulative or circulative-propagative) viruses are those which, when acquired by a vector, pass through the midgut wall to the salivary glands from where they can infect new hosts. Such viruses may multiply within tisues of a vector which retains the ability to transmit the virus for a considerable time, in some instances for the rest of its life. Persistent viruses, in contrast to those in the first two categories, may be quite specific with respect to the vectors capable of transmiting them (Matthews, 1970).

3.4 Insect Pests of Stored Products

Almost any stored material, whether of plant or animal origin, may be subject to attack by insects, especially species of Coleoptera (larvae and adults) and Lepidoptera (larvae only). Among the products that are frequently damaged are gains and their derivatives, beans, peas, nuts, fruit, meat, dairy products, leather, and woolen goods. In addition, wood and its products may be spoiled by termites or ants. Again the reader is referred to the appropriate chapters describing these groups for specific examples.

Borror *et al.*, (1976) cite a figure of about 1 billion dollars for the estimated value of losses caused by pests of stored products. The nature of the damage caused by these pests varies. Pests of household goods, including clothes, furniture, and various grain-based foods, cause damage as a result of spoilage, for example, by tunneling through or defecation on/in their food. Grain and other seed pests eat economically valuable quantities of food.

4. Pest Control

As will be apparent from what was said above, pests are organisms which damage, to an economically significant extent, Man or his possessions, or which in some other way are a source of "annoyance" to Man. Implicit in the above description are value judgements which may vary according to who is making them, as well as where and when they are being made

Nevertheless, in a given set of circumstances, there will be an economic injury (annoyance) threshold, measured in terms of a species' population density, above which it is desirable (profitable) to take control measures that will reduce the species' density. As the margin between economic injury threshold and actual population density widens, the desirability (profitability) of control increases. Pest control is, then, essentially a sociological problem—a matter of economics, politics, and psychology.

A range of methods is available for the control of insect pests. Each of these methods has its advantages and disadvantages and these must be balanced against each other in determining which (combination of) method(s) is most appropriate in a given instance. Some of these methods are spectacular but short-term and will be appropriate, for example, where massive outbreaks of pests are relatively sudden, yet unpredictable and temporary. Others are more slowly acting but relatively permanent in effect, and may be used for pests that are more or less permanent but whose populations are relatively stable.

Conway (1976) and Southwood (1977) have proposed that pests can be arranged in a spectrum according to their "ecological strategies" and that the principle (best) method of control is based on their position in the spectrum (Table 11.3). At either end of the spectrum are the so-called "r pests" and "K pests," respectively, with in between, the intermediate pests.

The r pests are characterized by their potentially high rates of population increase (due to the high fecundity and short generation time), well-developed powers of dispersal (migration) and ability to locate new food sources, and rather general food preferences. These features enable r pests to colonize temporarily suitable habitats, in which there is typically little interspecific competition for the resources available. Because r pests may occur in such large but unpredictable numbers and rapidly change their location, predators (of which there may be

TABLE 11.3

Principal Control Methods in Relation to Ecological Strategies of Pests*

	r Pests	*Intermediate pests*	*K Pests*
Control method	*Pesticides*——→	*Biological control*	←- - - -*Cultural control* ←- - - -*Genetic control*
Examples (with important features):	*Schistocerca gregaria* (desert locust) Fecundity/♀=400 eggs Generation time=1–2 months Migratory, defoliates many crops	(Most deciduous forest pests, fruit pests, and some vegetable pests)	*Oryctes rhinoceros* (rhinoceros beetle) Fecundity/♀=50 eggs Generation time=3–4 months Feeds on apical growing points of coconuts
	Aphis fabae (black bean aphid) Fecundity/♀=100 eggs Generation time=1–2 weeks Feeds on wide range of plants		*Glossina spp.* (tsetse fly) Fecundity/♀=10 eggs Generation time=2–3 months Feeds on narrow range of hosts
	Musca domestica (housefly)		*Corpocapsa pomonella* (codling moth)

Facundity/♀=500 eggs Generation time=2 – 3 weeks Feeds on organic waste	Fecundity/♀=40 eggs Generation time=2 – 6 months Larvae feed on apple and some other fruits
Agrotis ipsilon (black cutworm) Fecundity/♀=1500 eggs Generation time=1 – 1 1/2 months Feeds on seedlings of most crops	*Melophagus ovinus* (cheep ked) Fecundity/♀=15 eggs Generation time=1 – 2 months External parasite of sheep

* Data from Conway (1976) and Southwood (1977).

many) have relatively little effect on their population. Further, although like other organisms r pests are subject to disease, the latter is slow to take effect, by which time significant damage may have been done. Finally, because of their high reproductive potential, r pests are able to tolerate man, mortality and rapidly regenerate their original density. Hence, biological control, which is a relatively slow but long-term method, is of little use against r pests. For such pests specific insecticides, which can be stored for application at short notice, continue to be the most important tool in their control. Included in the r-pest group are the "classic" pests: locusts, aphids, mosquitoes, and housefiies (Table 11.3).

K pests, on the other hand, have lower facundity and longer generation time, poor ability to disperse, relatively specialized food preferences, and are found in habitats that remain stable over long periods of time. Under natural conditions, insects with the features of K strategists seldom become pests. If, however, probably as a result of Man's activity, their niche is expanded (*e.g.*, their food plant becomes an important crop), or if they can occupy a new niche (*e.g.*, feeding on domestic cattle rather than wild ungulates) they may become a pest. Once established, such pests are often difficult to eradicate over the short term, for example, through the use of insecticides. Insecticides are frequently not feasible tools because the K pests attack the fruit rather than the foliage of crop plants, or because the cost is prohibitive in view of the low density of the pest population. (In some instances, however, where even at low population density a pest may cause considerable damage, for example, codling moth on apple, insecticidal control may be profitable.) Nor is biological control an appropriate method because K pests have few natural enemies, a feature probably related to their low density under natural conditions. For K pests, the best methods of control are those which disturb a pest's habitat, for example, the breeding of resistant strains of plant(s) or animal(s) attacked by it, and cultural practices. Examples of K pests are given in Table 11.3.

The majority of pests are classified as intermediate pests in the Conway scheme because they exhibit a mixture of the features of r and K pests. For some of these, with a relatively high reproductive potential, insecticidal control may be necessary under certain conditions, and conversely, for pests approaching the K end of the spectrum, cultural control sometimes may be adequate. However, the most important feature of intermediate pests is the relatively large number of natural enemies that they have. These enemies, under normal circumstances, are important regulators of the pest population. In addition, intermediate pests are frequently foliage- or root-damaging pests, for example, spruce budworm and some scale insects, and, therefore, the economic injury threshold is reasonably high: that is, a fair amount of damage can be tolerated without economic loss. Hence, for these pests, biological control would appear to be the single most appropriate method of control, which can be supplemented as necessary with insecticidal and other methods. The latter is, in other words, an integrated control programme.

With these general considerations in mind, it is now appropriate to consider in more detail the methods available for pest control.

4.1. Legal Control

Also known as regulatory control, legal control is based primarily on the old adage "Prevention is better [in this instance, cheaper] than cure." Legal control is the enactment of legislation to prevent or control damage by insects. It includes, therefore, establishment of quarantine stations at major ports of entry into an area. Usually the stations are located at international borders, though in some instances domestic quarantines are necessary, for example, when certain parts of a country are widely separated from the rest (Hawaii and continental United States). At quarantine stations people and goods are inspected to prevent the accidental introduction of potential insect pests and plant and animal diseases. Prior to the introduction of quarantine legislation in the United States early this century (Plant Quarantine Act of 1912) a number of insect species had been accidentally introduced and become established as plant pests, for example, the cottony-cushion scale discussed in Section 2.3. Quarantine has severely reduced the number of pest introductions, though several species still managed to enter the United States, for example, the Mediterranean fruit fly (Ceratitis capitata) which became temporarily established in Florida and Texas on several occasions in the 1950s and 1960s, and the cereal leaf beetle (Oulema melanopus), found in Michigan in 1962 (Horn, 1976). The Mediterranean fruit fly was successfully exterminated using insecticides following each of its invasions, though at a cost of several million dollars. However, the cereal leaf beetle is now established in several midwestern states (primarily Michigan, Ohio, and Indiana) and efforts must now be directed toward restricting its dispersal.

Thus, another aspect of legal control is the setting up of surveillance systems for monitoring the insect population in a given area so that, should an outbreak occur, it can be dealt with before it has a chance to spread. Such surveillance is an important duty of state/provincial entomologists, in cooperation with local agriculture representatives and crop and livestock producers.

As an adjunct to quarantine, many countries (or areas within countries) have legislation that requires international or inter-state shipments of animals or plants, or their products, to be certified as disease- or insect-free by qualified personnel prior to shipment.

Another aspect of legal control, and one which is becoming increasingly important, is the licensing of insecticides and the establishment of (1) regulations regarding their use, and (2) monitoring systems to assess their total impact on the environment. In the United States the Environmental Protection Agency is responsible for assessing the effectiveness of pesticides, as well as their possible hazardous effects on humans, wildlife, and other organisms, including bees, other pollinating species, and beneficial parasitoids. As noted earlier, indiscriminate use of insecticides can result in greatly increased rather than decresed pest damage.

4.2. Chemical Control

The use of chemicals either to kill or to repel insect pests is the oldest method of pest control. Fronk (in Pfadt, 1971) notes that sulphur was used by the Greeks against pests almost 3000 years ago and the Romans used asphalt fumes to rid their vineyards of insect pests. The Chinese used arsenic compounds against garden pests before 900 A.D., though arsenic was not used in the Western world until the second half of the seventeenth century.

Until about 1940, insecticides belonged to two major categories, the "inorganics" and the "botanicals." Among the inorganic insecticides are arsenic and its derivatives (arsenicals), including Paris green (copper acetoarsenite), which was the first insecticide to be used on a large scale in the United States —against Colorado potato beetle in 1865. Other inorganic include fluoride salts (developed at about the end of the nineteenth century, following realization that toxic residues were left by arsenicals), sulphur, borax, phosphorus, mercury salts, and tartar. These inorganic insecticides were typically sprayed on the pest's food plant or mixed with a suitable bait. In other words, all are "stomach poisons" that require ingestion and absorption to be effective. Thus, they were unsatisfactory pesticides for sucking insects for which "contact poisons," absorbed through the integument or tracheal system are necessary.

The "botanicals" are organic contact poisons produced by certain plants in which they serve as protectants against insects. The group includes (1) nichotine alkaloids, derived from certain species of Nicotiana, including N. tabaca (tobacco) (family Solanaceae); (2) rotenoids extracted from the roots of derris (Derris spp.) and cube (Lonchocarpus spp.); and (3) pyrethroids, produced by plants in the genus Pyrethrum (Chrysanthemum) (family Compositae). The pyrethroids, when first available commercially, were an important group of insecticides for use in the home, as livestock sprays, and against stored-product, vegetable, or fruit pests, primarily because of their low toxicity to mammals. The great disadvantage or pyrethroids has been their photolabile nature (instability in light) and the need, therefore, to reapply them frequently made them expansive to use. Thus, they were largely replaced by cheaper, synthetic insecticides in the 1940s, an important consequence of which is that very few insects (6 of 364 species with insecticide resistance of some type) are presently resistant to them. This feature, in conjunction with the recent synthesis of photostable pyrethroids, may lead to renewed importance for this group of compounds (Elliott *et al.*, 1978).

Though two synthetic organic insecticides had been commercially available prior to 1939 (dinitrophenols in Germany, first used in 1892; organic thiocyanates in the United States from 1932 on), it is generally acknowledged that this is the date at which the synthetic organic insecticide industry took off. After several years of research for a better mothproofing compound. Muller, who worked for the Geigy company in Switzerland, discovered the value of DDT as an insecticide. In the next few years, production of DDT began at the company's plants in the United Kingdom and United States, although because of the second world war, knowledge of DDT was kept a closely guarded secret. In early 1944, DDT was first used on a large scale, in a delousing programme in Naples where typhus had recently broken out. Some 1.3 million civilians were treated with DDT, and within 3 weeks the epidemic was controlled, (Fronk, in Pfadt, 1971). Later that year the identity of the "miracle cure" was revealed, and the world soon became convinced that with DDT (and other recently developed insecticides) pest insects would become a thing of the past. In 1948 Muller was awarded a Nobel prize, though, interestingly, the first example of insect resistance to DDT had been reported two years earlier!

Through the 1940s and into the 1950s, much research was carried out in Western Europe and the United States for other insecticides as effective as DDT. Three major groups of compounds emerged:

1. chlorinated hydrocarbons, such as DDT, lindane, chlordane, aldrin, dieldrin, endrin, and heptachlor;
2. organophosphates, like TEPP, diazinon, dichlorvos, parathion, and malathion; and
3. carbamates, for example, sevin, isolan, and furadan.

For details of structure, physical properties, formulation, lethal doses, usage, etc., consult Pfadt (1971) and Metcalf *et al.*, (1962).

The search for suitable synthetic insecticides continues today, though at a somewhat reduced rate because the profitability of such ventures for industrial concerns has greatly diminished, for a variety of interrelated reasons. The primary reasons are (1) the time and cost of producing and testing an insecticide, estimated at an average of 7 years and more than 10 million dollars (Brown, 1977), with no guarantee of its final approval for use; (2) a general unwillingness of government agencies to grant approval for use of new insecticides following public outcry against the environmental hazards of such compounds; if approval is granted, it may restrict the use (and hence saleability and profitability) of an insecticide; and (3) the relatively short "life" of an insecticide due to its being no longer effective and/or becoming an environmental hazard. As a result, some of the largest companies in the chemical industry have either considerably reduced or abandoned research into the development of new insecticides (Brown, 1977).

Three major problems have arisen as a result of the massive use of insecticides over the past 70 years. First, many insects and mites have developed resistance to one or more of the chemicals (Tables 11.4 and 11.5, discussion of the mechanism of resistant.) Quadraspidiotus perniciosus (San Jose scale) takes credit for being in first recorded species to develop resistance to an insecticide (lime sulphur in 808, less -

than 30 years after use of the insecticide began. The housefly waire first species resistant to a synthetic insecticide (DDT in 1946), and by third of 1975 the total had reached 364, of which 139 species are of medical originary importance and 225 are agricultural pests (Georghiou and Taylor, 1937). Depending on the method of resistance developed by a species against insecticide, the species may also have resistance to other chemicals of the one group (class resistance) or even to chemicals of other groups (cross resistance). The development of resistance necessitates the use of ever-increasing loses of insecticide to achieve the same benefits with a concomitant ensurement of this second and third problems described below, as well as

TABLE 11.4

Number of Species of Insects and Mites in Which Resistance to One or More Chemicals has been Documented*

Year	*Species*
1908	1
1928	5
1938	7
1948	14
1954	25
1957	76
1960	137
1963	157
1965	185
1967	224
1975	364**

* Data from Georghiou and Taylor (1977).

** Includes 59 species reported on basis of field tests or observations; of these 3 are of medical veterinary importance and 56 of agricultural importance.

increased cost to the user. Alternatively, the user may turn to newly developed insecticides though again, the cost will be high.

The second problem associated with insecticide use issue already mentioned in Section 2.3, namely, the nonspecificity of action sense chemicals with the result that beneficial as well as pest species are delayed. (Indeed some were developed precisely because of their "general prime" nature!) The pest species typically can recover from insecticide applicant more rapidly than their natural enemies (because of their greater reproductive potential they rebound with even greater force, necessitating additional insecticides treatment and increasing costs to the user.

The third problem, the potential health hazard of many the synthesis insecticides to humans and wildlife, is of especial interest lanse it is base on a feature of the synthetics that was once considered highly beneficial, namely, their highly persistent (indestructible) nature. To the user this means that one or a few insecticide treatments will suffice for the entire season. For example, DDT is highly stable and only slowly degraded in the presence of sunlight and oxygen. Thus, a single spraying in a house or barn may remain effective up to a year, and even outdoor applications (on foliage) may be stable through an entire growing season. Unfortunately, this stability is retained following ingestion or absorption of DDT by living organisms. Thus, DDT tends to be stored in fatty tissue, due to its lipid solubility, and is concentrated as it is transferred from organism to organism in food webs. Recognition of the phenomenon of bioconcentration (biological magnification) via food webs and observation of harmful effects of insecticides in the terminal members of food chains (especially predatory birds but not, so far, humans) led to neormous public outcry against insecticides. As a result, governments have been forced to examine carefully the balance between the benefits gained and the risks entailed in the use of insecticides, and where necessary enact legislation to protect human interest. One result of this was the banning in 1972 of DDT use in the United States, in

TABLE 11.5

Number of Species of Arthropoda with Report loses of Resistance to Pesticides as of 1975*

	Pesticide group					Nature of pest		
	DDT	*Cyclo-dienes*	*Organo-phosphates*	*Carba-mates*	*Other*	*Medical veterinary*	*Agricultural*	*Total*
Acarina	21	10	32	6	13	10	33	43
Anoplura	5	3	2			5		5
Coleoptera	26	48	18	7	8		56	56
Dermaptera	1						1	1
Diptera	91	100	40	6	4	110	23	133
Ephemeroptera	2						2	2
Hemipt/Het.	4	12	3			4	10	14
Hemipt./Hom.	10	11	28	4	4		41	41
Hymenoptera	1	1					2	2
Lepidoptera	31	32	22	12	4		52	52
Mallophaga		2				2		2

Orthoptera	3	1	1	1		3		3
Siphonaptera	5	3	1			5		5
Thysanoptera	3	2			2		5	5
Total	203	225	147	36	35	139	225	364**

* Data from Georghiou and Taylor (1977).

** Includes 59 species reported on basis of field tests or observations: of these 3 are of Medical Veterinary importance and 56 of agricultural importance. Note that many species are resistant to pesticides or more groups.

all except a very few situations where benefits clearly outweighed risks (Whittemore, 1977).

Despite this rather gloomy picture, it must be strongly emphasized that synthetic insecticides have saved and will continue to save millions of human lives and billions of dollars' worth of food and organic manufactured goods. They are still by far the principal method for control of insect pests, though there is a slowly growing realization among users that insecticides are probably more valuable (*i.e.*, cheaper and having a longer period of service) when used selectively, in conjunction with other methods such as biological control, rather than in a "blanket" manner as in the past (see Doutt and Smith, in Huffaker, 1971). This is associated with an appreciation that actual extermination of a pest is almost never achievable, or is even necessary in most situations; that is, a certain amount of pest damage can be tolerated without suffering economic loss.

The chemicals discussed above operate on the principle of control through rapid death of (most) members of a pest population. Recently, however, great interest has been shown in other chemicals which, though widely different in nature, are collectively known as insectistatics because they suppress insects' growth and reproduction (Levinson, 1975). They include (1) substances that inhibit chitin synthesis or tanning, rendering an insect more susceptible to microbial (especially fungal) infection or preventing normal activity because the muscles do not have a firm structural base; (2) antagonists or analogues of essential metabolites (*e.g.*, essential amino acids and vitamins) whose effect is to prolong larval life and or retard egg production, frequently resulting in death; (3) insect growth regulators (IGRs), substances with juvenile-hormone or ecdysonelike activity; and (4) sex attractants.

Some 600 compounds are known to mimic juvenile hormone to various degrees and in different species. Work on the possible use of IGRs in pest control is still largely at the stage of laboratory experiment or small-scale field trial though early results indicate considerable promise for IGRs against pests of some orders (Staal, 1975). Among the effecrs noted are (1) interference with embryogenesis, followed by death, at IGR does about 1000th the value of conventional ovicides; (2) abnormal development of the integument in post-embryonic states, leading to inability to molt properly andimpaired sensory function (hence inability to locate food, mates, oviposition sites, etc.); (3) improper metamorphosis of internal organs or external genitalia, causing sterility and/or inability to mate; (4) interference with diapause, so that an insect becomes seasonally maladujusted; and (5) abnormal polymorphism in aphids. Despite the seemingly bright picture for IGRs at the present time, they might, if used indiscriminately, soon become subject to the same criticisms that have been levelled at conventional insecticides.

Another area of chemical control that holds considerable promise is the use of sex attractants. Considerable progress has been made in the elucidation of their structure and artificial synthesis. Though the substances exert their effect at extremely low concentrations, the high cost of manufacture and technical problems associated with their application presently prohibit large-scale use. However, it may be anticipated that pheromones may find the following uses in pest control:

1. Estimation of population density (following by treatment, if necessary, with a suitable control methed);
2. Attraction of pests into traps in which they are then killed; and
3. Permeation of an environment with sex attractant, so that individuals are unable to locate mates.

Biological Control

Biological control, in the sense used here, may be described as the regulation of pest populations by natural enemies (parasites, predators; and pathogens). It includes both naturally occurring control and control achieved as a result of Man's augmentation of the natural enemy component (through importation, rearing and release of selected enemies).

As in the history of chemical control, several early successes with biological control led some scientists to believe that this method might be the one to solve the world's pest problems. However, with an increase in knowledge of ecology, specifically predator-prey and host-parasite relationships, and in the length of the list of "failures" in biological control projects, this view has been markedly tempered. It is now realized that for some pest species, biological control will not work, and that for some others, biological control has been or will be a highly effective method. Between these two extremes, and forming the majority of pests, are those for which biological control will be an effective tool when used in conjunction with other methods of pest control that is, as a component of integrated pest control. For many years, biological control was a "poor relation" to chemical control, even though several outstanding successes of biological control were recorded before the advent of synthetic insecticides. However, with the increasing appreciation of the problems causes by synthetic insecticides, biological control began to receive a greater share of the attention of applied entomologists, industrial concerns, and government agencies.

The importance of naturally occurring biological control most be emphasized. DeBach (1974) suggests that "Upon it rests our entire ability to successfully grow crops because without it, the potential pests would overwhelm us." He estimates that "99 per cent or more of potential pests (are) under natural biological control." However, by its very effect, namely, the prevention of species from becoming sufficiently populous to be designated as pests, it is easily overlooked. Only by very careful study of ecosystems in equilibrium or by disturbance of ecosystems can its value be appreciated. Two common (man-made) methods of ecosystem disruption are (1) transfer of a species from its original habitat where it is not a pest to a new habitat where, in the absence of natural enemies, it flourishes and becomes a pest, and (2) indiscriminate use of broad-spectrum insecticides which decimate both pest and natural enemy populations. As noted above, the pest normally recovers more rapidly than its predators or parasitoids and becomes even more destructive than before. Examples of both forms of disruption were given in Section 2.3. DeBach (1974) lists 20 cases of naturally occurring control of homopteran pests (mostly scale insects), and further examples involving pest insects of other orders are given in the papers by Hagen *et al.*, Rabb, and MacPhee and MacLellan, in Huffaker (1971).

Man's first recorded attempts at biological control were made by the ancient Chinese who used predatory ants in their citrus orchards against caterpillars and boring beetles. This practice is still used in parts of Burma and perhaps China (DeBach, 1974). In the eighteenth century, mynah birds imported from India were successfully used in the control of red locusts (Nomadacris septemfasciata) in Mauritius. However, full appreciation of the potential that insect parasitoids and predators might have as control agents developed only in the middle of the nineteenth century, following careful studies of the biology of such insects in the early 1800s. During the latter half of the nineteenth century, many well-known entomologists, both in Europe and North America, studied biological control and extolled its virtues, though these were largely unheeded until cottony-cushion scale was spectacularly controlled by the vedalia beetle in 1888-1889. More or less concurrently, based on studies of insect diseases (especially those of the silkworm), several authorities, most notably Pasteur and an American entomologist LeConte, suggested that microorganisms might be used in the control of pest species (DeBach, 1974). However, for reasons outlined below, successful biological control through the use of microorganisms has been achieved relatively infrequently.

Biological control, when successful, has several advantages over control though the use of insecticides. First, it is persistent; that is once a control agent is established, it will exert a continuing influence on the population density of the pest. In part related to this, biological control is cheap because one application of the control agent is sufficient. Furthermore, the control agent is "ready-made"—it does not have to go through an extended and costly phase of research and development, though determination of the most suitable control agent(s) may take some time. As an example of the cheapness of biological control, DeBach (1974) notes that the cottony-cushion scale project in 1888-1890 cost less than 5000 dollars, yet it has saved the California citrusfruit industry millions of dollars each year since. Because of escalating costs of insecticides, biological control may be the only (or principal) method of pest control in underdeveloped countries. A third advantage is that biological control does not stimulate "genetic counter-attack" by the pests. Fourth, it does not result in the growth to pest status of species that are economically unimportant. And finally, it is a method that does not endanger humans or wildlife through pollution of the environment (Huffaker, 1971).

If biological control offers so many advantages over insecticides, why has it, in most instances, come only a distant second to these compounds in terms of usage? What are its disadvantages? Perhaps the main one is psychological because users like to see immediate results (profits?) for their efforts. In biological control it may take some time (even years) for a control agent to subdue a pest, by which time a user's patience (and profit margin) have worn thin, especially under the considerable and continuous advertising pressure of insecticide producers. Further, the new equilibrium density that a pest attains when controlled biologically is almost certainly higher than the density immediately after insecticide treatment, which again may be unsatisfactory so far as a user is concerned, especially if the crop being grown is of high unit value, for example, fresh. fruit. Consumers, too, are involved here, since they have become accustomed to "blemish-free" produce and may not buy even slightly damaged material.

Biological control has not always been successful, and proponents of insecticidal control (which include a significant number of applied entomologists as well as industrial concerns) have been quick to point this out. As a result many governments and universities have been loath to invest money and manpower in research on biological control. According to DeBach (1974), had such research been possible, a number of major pests could have been controlled (biologically) long before they were. In the few countries or states (most notably. California) where significant research effort has been put into biological control, the success rate for this method is high. In other words, in many instances, failure to control a pest by biological measures has stemmed from incomplete or poorly performed study of the pest and its predators or parasites, not from the unsuitability of biological control per se.

The greatest success has been achieved through the use of insects, especially parasitoids, as control agents and some examples are given in Table 11.6. According to DeBach (1974), up to 1970 biological control using insects as control agents had been attempted against 223 pest species throughout the world. Forty-two of these species were completely controlled; that is, their population densities were reduced, then permanently maintained, below the economic threshold, making insecticide treatment only rarely or never necessary. Substantial control was achieved over 48 species. Substantial control includes instances where the pest or crop is not of major importance, where control is only partial in some areas of infestation and where occasional insecticide treatment is required. Another 30 species fell into the "partial control" category, where pest density was reduced but not consistently below the economic threshold, making regular, though less frequent, insecticidal treatment necessary. In short, some degree of success was achieved against more than one half of the species against which the method was used. Indeed, the total number of successful projects up to 1970 was 253 on a world basis, since some of the pests were controlled in several different countries.

Not surprisingly, in view of the "unprofitability" of successful (long-term) biological control using insect agents, industry has shown little interest in the method. Research and development in this area have been the domains of government and university scientists.

TABLE 11.6

Examples of Successful Biological Control Projects Using Insects as Control Agents*

Pest	*Primary control agent*** *and source*	*Location and date of project*
1	*2*	*3*
Iceryo purchasi (cottony-cushion scale)	Vedalia cardinalis (vedalia beetle? (Australia)	California (1888-1889)
Perkinsiella saccharcida (sugarcane leathopper)	Paranagrus optabilis (mymarid) (Australia) Cytorhinus mundulus (mirid) (Australia)	Hawaii (1904-1920)
Levuana irridescens (coconut moth)	Ptychomyia remota (tachinid) (Malaysia)	Fiji (1925)
Aleurocanthus woglumi (citrus blackfly)	Eretmocerus serius (aphelinid) (Malaysia)	Cuba (1930), other Caribbean islands and Central American countries (1931 on)
	Amitus hesperidum (platygasterid) Prospaltella opulenta (aphelinid) P. clypealis (aphelinid) (India and Pakistan)	Mexico (1949)
Planococcus kenyae (coffee mealybug)	Anagyrus nr. kivuensis (encyrtid) (Uganda)	Kenya (1938)
Pseudococcus citriculus (citriculus mealybug)	Clausenia purpurea (encyrtid) (Japan)	Israel (1939-1940)
Dacus dersalis (oriental fruit fly)	Opius oophilus (braconid) (Philippines and Malaysia)	Hawaii (1947-1951)

1	*2*	*3*
Lepidosaphes beckii (purple scale)	Aphytis lepidosaphes (eulophid) (China)	California (1948 on). Texas, Mexico, Greece, Brazil, Peru (1952-1968)
Antonina graminis (rhodesgrass scale)	Neodusmetia sangwani (encyrtid) (India)	Texas, Florida, Brazil (1949 on)
Operophtera brumata (winter moth)	Cyzenis albicans (tachinid) (Europe) Agrypon flaveolatum (ichneumon) (Europe)	Nova Scotia (1955-1960)
Chrysomphalus aonidum (Florida redscale)	Aphytis holoxanthus (eulophid) (Hong Kong)	Israel, Lebanon, Florida, Mexico, South Africa, Brazil, Peru (1956 on)
Chrysomphalus dictyospermi (dictyospermum scale)	Aphytis melinus (eulophid) (California)	Greece (1962)
Nezara viridula (green vegetable bug)	Trissolcus basalis (scelionid) (Egypt and Pakistan)	Australia (1933-1962), New Zealand (1949-1952). Fill and other Pacific islands (1941-1953)
	T. basalis and Trichopoda pennipes var, pilipes (tachinid) (Antigua and Monserrat)	Hawaii (1962-1965)
Opuntia spp. (prickly pear cactus)	Cactoblastis cactorum (moth) (Argentina)	Australia (1920-1925)
Hypericum perforatum (Klamath weed)	Chrysolina quadrigemina (chrysomelid beetle) (Australia)	California (1944-1946) and other western states

*Data from DeBach (1974)

**In many projects, a variety of control agents were introduced and became established, but assist in control of the pest to only a minor degree.

Biological control using microorganisms (microbial control), specifically viruses, bacteria, protozoa, fungi, and nematodes has not met with an equivalent degree of success (see Burges and Hussey, 1971; Bulla, 1973; Angus 1977). This stems largely from the nature of the control agents, many of which are, obligate pathogens, relatively labile (short-lived) outside the host, have poor powers of dispersal, and are active only in certain environmental conditions. Added to these problems, are technical difficulties that have hampered research in this area, especially the inability to mass-produce microorganisms on a year-round basis for laboratory study as well as field trials. This is slowly being overcome through the use of artificial insect diets, enabling cultures of the host insect to be maintained year-round and tissue culture. However, these methods are not, by and large, satisfactory for large-scale commercial production because of their labour-intensive nature and high costs. Another major technical problem that requires solution is the development of suitable protectants against sunlight to which most microorganisms are especially sensitive. The poor viability of most microorganisms outside their host means that, to be effective, they must be reintroduced at least on an annual basis, if not more frequently. In this regard they are, of course, comparable with chemical insecticides, though they do not pose the same environmental hazards. The continuing need for reapplication of microbial agents has not gone unnoticed by industrial concerns, and a number of commercial preparations are now available or under development (Table 11.7).

Four examples are known of microorganisms that exert a permanent regulatory effect on pest insects. These are Bacillus popilliae and B. lentimorbus, which cause milky disease in the Japanese beetle, Popilla japonica, an important pest of lawn and other grasses in the United States; a nuclear polyhedrosis virus, accidentally introduced into Canada in the early 1940s, which has kept populations of the European spruce sawfly (Diprion hercyniae) below the economic injury threshold for more than 30 years since the initial epizootic; and a nuclear polyhedrosis virus which exerts good control over the European pine sawfly (Neodiprion sertifer) in Canada Falcon, in (Huffaker, 1971).

Bacillus thuringiensis, which, as noted in the previous chapter, is pathogenic to a large number of Lepidoptera, has been available commercially in the United States since 1958. As of 1971 it was the only microbial agent registered for use against crop, as well as forest and shade-tree, pests. The great advantage of B, thuringiensis is that it can be cultured outside its hosts by fermentation and is therefore cheap to produce. The bacterium is short-lived, however, and several applications may be necessary each season.

As Table 11.7 indicates, a number of viruses (especially nuclear polyhedrosis viruses) have been developed for use against insect, especially lepidopteran, pests, though only a few of these have been registered for commercial use. Two major problems beset the use of viruses as microbial control agents, their low viability in sunlight and their culture. Although in soil viruses may survive from season to season, when exposed to ultraviolet light, their half-life is usually less than 1 day. Even when additives which screen out ultraviolet light are added to the preparation, the half-life is only extended to a few days. As of 1975 all viruses used in field trials or available commercially were produced in living insects, the mass-rearing of which may be time-consuming and expensive. This difficulty may eventually be overcome through the use of tissue culture, though at the moment this method is also relatively costly.

It was a fungus (Metarrhizium anisopliae) that was the control agent in the first attempt at microbial control. Generally speaking, however, the potential of fungi as pest control agents has not been widely studied. Roberts (in Bulla, 1973) suggests three reasons why workers may have been unwillng to commit themselves to such a study. First, fungi appear to be unpredictable; that is, in some tests high infectivity of hosts was achieved, whereas in others the opposite occurred. Roberts considers this inconsistency to stem from inadequate information on the fungal requirements for infectivity. Second, because

TABLE 11.7

Examples of Insect Pathogens Developed by Industry and Various Agencies*

Group	*Pathogen*	*Product Name*	*Source*
Bacteria	Bacillus lentimorbus	Japidemic	Ditman Corp., U.S.A.
	B. papilliae	Doom	Fairfax Biological Labs., U.S.A.
	B. sphoericus	—	International Minerals Chemical Corp. (IMC), U.S.A.
	B. thuringien	Agritrol	Merck and Co., U.S.A.
		Bakthane L69	Rohm and Hass Co., U.S.A.
		Bactospeine	Pechiney Progil Lab. Roger Bellon, France
		Bathurin	Chemapol, Biokrma, Czechoslovakia
		Bickpor 2802	Farbwerke Hoechst, Germany
		Biotrol BTB	Nutrilite Products, Inc., U.S.A.
		Dendrobacillin	Moskovs. zavod. bakt, prepavatov. atov., U.S.S.R.
		Entobakterin 3	All-Union Institute Plant Protection, U.S.S.R.
		HD-1	Abbott Laboratories, U.S.A.

		(Experimental) Parasporin	Grain Processing Corp., U.S.A.
		Sporeine	Laboratoire L.I.B.E.C., France
		Thuricide	IMC, U.S.A.
Fungi	Beauveria bassiana	—	IMC, U.S.A.
		Biotrol FBB	Nutrilite Products, Inc., U.S.A.
	Metarrnizium anisopliae	—	IMC, U.S.A.
Polyhedrosis viruses	Hellothis	Biotrol VHZ	Nutrilite Products, Inc., U.S.A.
		Viron/H	IMC, U.S.A.
	Neodiption	Polyvirocide	Indiana Farm Bureau Co-op Assoc., U.S.A.
	Prodenia	Biotrol VPO	Nutrilite Products, Inc., U.S.A.
	Spodoptera	—	IMC, U.S.A.
		Biotrol VSE	Nutrilite Products, Inc., U.S.A.
	Trichoplusia	—	Biological Control Supplies, U.S.A.
		—	IMC. U.S.A.
		Biotrol VTN	Nutrilite Products, Inc., U.S.A.
Nematodes	Neoaplectana carpecapsae (DD-136)	Biotrol NCS	Nutrilite Products, Inc., U.S.A.

*Data from Falcon (in Huffaker, 1971).

fungal infections depend heavily on weather conditions, which cannot be controlled, the use of fungi as control agents was thought to be impractical. However, it is the microclimate (rather than weather conditions) which is important, and this is amenable to modification under certain circumstances. Third, because some fungi that are pathogenic in insects are also known to affect vertebrates, the tendency has been to ignore the group as a whole on the grounds of safety.

Nonetheless, fungi have same qualities that indicate potential as control agents, for example, many (even some which were once thought to be obligate parasites) can be cultured on artificial media, they may be highly virulent and among different species a range of host specificity can be found, so that a species appropriate for a given control problem may be selected (Roberts, in Bulla, 1973, see also Ferron, 1978).

Very little work has been done on the possibility of using pathogenic protozoa in the biological control of insect pests, though some, especially species of Microsporidia, are probably important natural control agents Kellen, 1974. As with other microorganisms, a serious deterrent has been the inability to culture the protozoa outside the host. However, using spores produced in mass-reared hosts, a number of Nosema spp. have been field tested with good results. For example, an extensive study is being undertaken in the United States by Henry and colleagues (see Henry and Oma, 1974) in the potential use of N. locustae, N. acridophaga, and N. cuneatum as control agents against grasshoppers. Using N. locustae spores mixed with bran bait, Henry and co-workers obtained significant reductions in the grasshopper populations of experimental plots However, the timing of spore application was found to be critical. Application too early in the season, that is, when grasshoppers were in the early larval instars, caused rapid and high mortality, with the result that too few spores were produced for infection of other grasshoppers. Conversely, application to older (near adult) grasshoppers, while resulting in a high incidence of infection (most individuals were infected), caused little mortality because the grasshoppers were able to tolerate the protozoa. (It is probable, however, that the fecundity of these individuals would be reduced). The best time of application appeared to be when the dominant, early summer species were in the third instar. Such applications. A caused the greatest initial mortality and highest infection in later populations. A comparable programme to that intiated by Henry's group has recently begun in Western Canada.

Like protozoa, pathogenic nematodes have been little studied with regard to their potential as control agents, largely because of the difficulty of culturing them on artificial media (Dutky, 1974; Pramer and Al-Rabai, in Bulla, 1973). As of 1974 only two species of nematodes pathogenic in insects [Neoaplectana glaseri and N. carpocapsae (=dutkyi)] could be cultured outside their host. Neoaplectana carpocapsae (also known as DD-136) appears to be a good candidate for a biological control agent. Its advantages, apart from ease of culture, include a high reproductive capacity, a potentially long-lived infective stage, and resistance of the infective stage to many insecticides (making it possible to use the nematode in conjunction with conventional chemical methods of pest control). Another possible advantage is its wide host range, which includes many pests, though this must be weighed against its detrimental effects on some beneficial insects. Its disadvantages include the need for a moist surface in order to migrate in search of a host and its sensitivity to high temperatures and desication. Hence, the timing of field applications is very important (Dutky, 1974).

4.4. Genetic Control

Methods for genetic control fall into two distinct categories:

1. those by which pests are rendered less capable of reproduction, and
2. those in which resistance is increased in the organism attacked by the pest.

A variety of genetic methods are potentially applicable for regulation of a pest's reproductive capability [see reviews by Proverbs (1969), and Whitten and Foster 1975)]. To date, however, only one of these, the sterile insect release method (SIRM), has been used on a full scale; the rest are either under examination in laboratories or in field trials or still at the theoretical stage. Knipling (1955) first proposed the idea of releasing sterile insects into wild populations of the pest so as to reduce total fecundity. Successive releases of sterile insects over a number of generations would lead, cumulatively, to eradication of the pest. SIRM has been used with striking success against the screwworm fly (Cochliomyia hominivorax) in Curacao and the southeastern United States where the fly has been eliminated, and in the southwestern United States where good control has been achieved, though not eradication because of immigration of wild flies from Mexico (DeBach, 1974). Flies are mass-cultured and sterilized by irradiation. SIRM has also been used successfully against small, isolated populations of other pests, for example, melon fly (Dacus cucurbitae) and Oriental fruit fly (D. dorsalis) in the Mariama Islands. Field tests have indicated that the method holds promise for the control of codling moth (Carpocapsa pomonella) in the United States and Canada, Mediterranean fruit fly (Ceratities capitata) in Spain, Italy, Hawaii, and Nicaragua, and Queensland fruit fly (D. tryoni) in Australia. The advantages of SIRM are its specificity, the permanency of its effect (though it may take several years to achieve this), and the fact that it does not pollute the environment. An important disadvantage is its limited applicability. It is not a very suitable method for species that mate frequently as the chances of successful (fertile) matings are increased. In this regard, it is worth noting that the pets controlled by this method are mostly Diptera or Lepidoptera, in many species of which females mate only once or a very few times. Nor is it feasible to use SIRM on pests that appear sporadically or in high density; the latter would make it very difficult to achieve the necessary high ratio of sterile: wild males in the field. In addition, there are several technical problems related to the mass production of individuals all of the same sex. Presently, both males and females reared in culture are irradiated and released. Thus, irradiated females will compere with wild females for the males attention, reducing the efficacy of the method. Two solutions have been proposed for this dilemma:

1. Incorporation of lethal sex-linked mutants into laboratory cultures. Such mutants might be, for example, temperature-sensitive, so that exposure of the population to a certain temperature would kill one sex, or sensitive to a particular insecticide. Thus, treatment of juvenile stages with the insecticide would leave only the resistant sex alive.
2. Chemosterilization of wild populations. This probably would be cheapter than mass-rearing followed by irradiation, though this would have to be weighted against the potential nonspecificity and pollution of the environment by the sterilant.

The use of irradiation or chemosterilants to achieve sterility (known as induced sterility) may not always be feasible due to health hazards, costs, etc., and a number of workers are now examining various methods of causing inherited sterility. This is sterility which results either from matings between genetically or cytoplasmically incompatible partners, with production of inviable zygotes, or because gametogenesis is abnormal in one or both sexes. For further details Whitten and Foster (1975) should be consulted.

The development of plant varieties capable of resisting attack by pests is a well-known method of pest control. Probably the classic example of plant resistance is the resistance of wheat varieties to attack by the Hessian fly Mayetiola destructor). Resistance of "Underhill" wheat was recorded in New York in 1782 (Gallun *et al.*, 1975), and several resistant vareities are now used regularly in the United States, with enormous annual savings.

Basically, increased resistance can be achieved by introducing either physical or biochemical changes in plants. For example, wheat resistant to attack by wheat stem sawfly (Cephus cinctus has solid rather than hollow stems; wheat grown in Michigan and resistant to cereal leaf beetle (Oulema melanopus) has "hairier" leaves than other varieties; resistance to Hessian fly is due to the presence of antibiotics in the plant tissues which causes the death of the larvae. Other examples of plants having varieties resistant to pests are corn to the European corn border (Ostrinia nubialis), and corn earworm (Heliothis zea); alfalfa to the pea aphid (Acyrthosiphon pisum); cabbage to the potato leafhopper (Empasca fabae), and beans to the Mexican bean beetle (Epilachna varivestis). All told plant varieties resistant to more than 100 pest species are known (DeBach, 1974).

The advantages of using resistant varieties include greater crop yield, relatively low cost of development, and absence of side effects (environmental damage, etc.). In addition, pests that feed on these plants may be more susceptible to disease, adverse weather conditions, and insecticides which can therefore be used in lesser amounts. Disadvantages include the length of time required to develop a resistant variety, ordinarily 10 to 15 years with crop plants and even longer with trees. Related to this, an extensive programme of screening must be carried out to ensure that the varieties are totally satisfactory. Thus, resistance may vary according to the form of the insect, for example, between apterous and winged aphids. Or increased resistance to one pest may result in decreased resistance to another. Another problem may be psychological, that is, to persuade a grower that a new variety is superior to the one previously used.

4.5. *Cultural Control*

Cultural control, the use of various agricultural practices to make a habitat less suitable for reproduction and/or survival of pests, is a long-established method of pest control. Cultural control aims, therefore, to reduce rather than eradicate pest populations and is typically used in conjunction with other control methods. However, in some instances, cultural practices alone may effect almost complete control of a pest, as occurs with the tobacco hornworm (Manduca sexta) in North Carolina and the pink bollworm (Pectinophora gossypiella) on cotton in central Texas.

The agricultural practices used either may have a direct effect on the pest or may act indirectly by stimulating population buildup of a pest's predators or parasites, or by making plants more tolerant of pest attack. An essential prerequisite for effective cultural control is detailed knowledge of a pest's life history so that its most susceptible stages can be determined. Important agricultural practices include (1) crop rotation to prevent buildup of pest populations; (2) planting or harvesting out of phase with a pest's injurious stage(s), which is especially important against species that have a limited period of infestation or for plants with a short period of susceptibility; (3) use of trap crops on which a pest will concentrate, making its subsequent destruction easy; (4) soil preparation, so as to bury or expose a pest, or increase the crop's strength so that it can more easily tolerate a pest; and (5) clean culture, the removal, destruction, or ploughing under of crop remains, in or under which pests may hibernate.

4.6. Integrated Control

Integrated control, synonymous with pest management, is "a pest population management system that utilizes all suitable techniques either to reduce pest populations and maintain them at levels below those causing economic injury, or to so manipulate the populations that they are prevented from causing such injury." It is an approach to pest control that arose, of necessity, with the realization that, for almost all pests, existing methods either individually were not capable of exerting permanent control, gave rise to harmful side effects, or created, by their very use, new pest problems.

Integrated control is not a new concept. It was envisioned by C.V. Riley, former Chief Entomologist in the United States, in the 1980s and was practised in the 1920s in California citrus groves (DeBach, 1974). However, it was not in most instances given serious consideration until the use of insecticides alone had increased to the point at which an entire crop-growing operation became unprofitable. (The importance of insecticides as environmental hazards did not enter into consideration until relatively recently). As has been noted on previous occasions, a major fault of pesticide use has been its indiscriminate nature. Insecticides were not used only as necessary (a decision based on pest population density) but on a regular (calendar) basis as a form of "preventive medicine" or "insurance," This approach to the use of pesticides, referred to by Doutt and Smith (in Huffaker, 1971) as "the pesticide syndrome," led rapidly to the development of resistance in the pests against which it was aimed, to the use of larger doses or of different insecticides (to which also the pests soon became resistant), to a rise to pest status of previously innocuous species, and ultimately to greatly elevated costs and decreased profits.

The primary step toward integrated control is collation of as much information as possible about the agroecosystem being studied; the more information that is available, the closer an integrated control programme will come to providing maximum returns from the investment of time, effort, and money. In many early integrated control projects, relatively little information was available on which to base integrated control programmes, and these were developed largely by intuition, despite which they were remarkably successful. Nowadays the considerably greater body of information generally available is stored in computers for use in constructing and testing mathematical models of the agroecosystem. Once developed, such models can be used to examine the effects of varying one or more of the factors that influence the population density of a pest and, therefore, to determine the optimum method of reducing and maintaining this density below the economic injury threshold.

Some of the important questions to be answered in an integrated control programme might be as follows. (1) What is the economic injury threshold? It has been observed frequently that insecticide users were maintaining pest population density at levels much below the economic injury threshold, increasing their operational costs in both the short and the long term. (2) What are the best parasitoids and predators to use and how might their abundance be increased? (3) What are the best insecticides to use, when should they be applied, and what are suitable doses? To answer this question requires consideration of the specificity of action, potential side effects, and stability of the insecticides. (4) Are suitable cultural procedures available? (5) Are resistant strains of the crop available? It will be apparent from the questions that pest management is interdisciplinary in nature and may require collaboration between experts in widely different areas, for example, economists, statisticians, meteorologists, plant breeders, entomologists, chemists, and toxicologists.

Many examples of pest management might be cited, that differ in complexity, geographic location, and crop pest being controlled. In its simplest (and rarest) form, pest management involves only one method of control and, strictly speaking, cannot be described as "Integrated control." Thus, biological control is entirely satisfactory for control of sugarcane pests in Hawaii and coconut pests in Fiji (DeBach, 1974). More often, two or more control methods are employed. For example, at the height of the pesticide syndrome, apple growers in Nova Scotia used a battery of fungicides and insecticides (especially broadspectrum synthetics) against apple pests. Far from achieving the desired effects, this led to even greater problems. By about 1950 earlier pests such as codling moth (Carpocapsa pomonella), eye-spotted bud moth (Spilonota ocellana) and oystershell scale (Lepidosaphes ulmi) had become even more serious, and other species, for example, European red mite (Panonychus ulmi) and fruit-tree leafroller (Archips argyrospilus) had become pests due to destruction of their natural enemies. The need to reduce the escalating costs of this programme stimulated intensive study of apple orchard ecology, including the effects of pesticides on the natural enemy complex. The outcome of this work was the development of an integrated control programme which utilized fewer but more specific insecticides (as well as different fungicides, some of which had killed some natural enemies) and allowed for recovery of many of the natural enemies. By 1955 the mite and scale insect had practically disappeared, and the density of the eye-spotted bud moth had fallen significantly.

As a final example, the integrated control programme for cotton pests in the San Joaquin Valley, California will be outlined (van den Bosch *et. al.*, in Huffaker, 1971). The usual picture emerges. After extensive use of organochlorine insecticides, followed by organophosphates and carbamates, there was a population resurgence and an increase in resistance of the two major pests, the corn earworm (Heliothis zea) and a mirid bug (Lygus hesperus), as well as outbreaks of secondary pests such as cabbage looper (Trichoplusia ni), beet armyworm (Spodoptera exigua), salt-marsh caterpillar (Estigmene acraea), and spider mites (Tetranychidae).

Early in the development of the integrated programme, it was realized that no reliably established economic injury thresholds existed for either of the major pests and that, for Lygus, time of insecticide application was critical. For Lygus, it was determined that the generally accepted injury threshold of 10 bugs/50 net sweeps was invalid; rather, the major period of importance was June 1 to July 20 and only if the density exceeded 10 bugs/50 net sweeps in this perion should insecticides be applied. After July 20, Lygus densities of 20 or more bugs 50 net sweeps would not affect the quantity or quality of cotton produced, provided that the plants had flowered normally June and early July.

It was noted that corn earworms seldom became pests except in fields treated for Lygus outbreaks (using the old value for economic injury threshold) where the corn earworms' natural enemies (the bugs Geocoris pallens and Nabis americoferus and the lacewing, Chrysopa carnea) were destroyed by insecticide. With the new threshold for economic injury due to Lygus, tied in with the plant's seasonal development, the use of insecticides was reduced, especially after late July, and the earworms' natural enemies survived. Further reduction in insecticide usage came about with the raising of the economic injury threshold for corn earworms from four earworms plant to 15 treatable earworms (first and second instar larvae)/plant. This arose, in part, from realization that larvae often feed on surplus buds, flowers, and small bolls that are not picked.

Cultural control is also used to reduce the effects of Lygus on the cotton crop. Lygus prefers alfalfa (Medicago sativa) to cotton and will remain on the former plant when given a choice. Traditionally, however, alfalfa fields are solid cut (all the field is cut at once) provoking mass migration of Lygus into adjacent cotton fields. Alfalfa growers are now being persuaded to practise strip cutting to prevent such migration, though the method is not widely adopted. Probably a better technique is for cotton growers to plant strips of alfalfa through their cotton fields, on which Lygus will concentrate.

Another cultural method that may become commercially feasible is the use of food sprays in attempts to prolong, the seasonal effectiveness of natural enemies whose populations frequently decline after midseason. For example, in experimental plots adult Chrysopa were attracted to the food sprays and produced more eggs, resulting in significantly less injury to plants by corn earworms (see Hagen and Hale, 1974).

Finally, microbial control may become an important method for use against the corn earworms and other lepidopteran pests of cotton. Some growers use homemade sprays of cabbage looper nuclear polyhedrosis virus, prepared from infected larvae which they collect. Heliothis nuclear polyhedrosis virus is now registered for use on cotton. Bacillus rhuringiensis is also registered for commercial use but is expensive and not sufficiently virulent against corn earworm to merit widespread application.

Integrated control results in substantial savings, both financial and in terms of pollution, through the greatly decreased use of insecticides. For example, in the San Joaquin Valley, the traditional control programme for cotton pests cost about twice as much per acre as the integrated programme, and even greater savings may be incurred as the integrated programme is improved. Adkisson (1971, cited in DeBach, 1974) considers that insecticide use against cotton pests could be reduced by 50 per cent throughout the United States without loss of production. As more than one half of the insecticide produced in the United States (totalling about 0.5 billion kilograms in 1971) is used on cotton, enormous savings are possible in the growing of this crop alone. Indeed, a 50—80 per cent reduction in the use of insecticides, after the initiation of an integrated control programme, has been observed in many other instances. Despite these remarkable savings, a great majority of crop growers and government officials remain convinced that the present method of pest control, the unilateral use of insecticides, is best. Persuading these individuals that to continue with this approach will lead not only to financial ruin but to long-term, perhaps irreversible, deterioration of environmental quality is perhaps the greatest challenge that scientists have ever faced.

5. Summary

Only a small proportion (perhaps 0.5%) of insect species interact directly or indirectly with Man. Most of these are beneficial to Man's existence by virtue of their pollinating activity, the materials they produce, or their role as biological control agents. Relatively few insects (about 0.1% of all species) constitute pests which alone ot in conjunction with microorganisms cause damage of death to Man, crops, livestock, and manufactured goods.

Pest insects may be classified according to their ecological strategies as r pests, K pests, or intermediate pests. The r pests have potentially high rates of population increase, well-developed ability to disperse and find new food sources, and general food preferences. Though they have a large number of natural enemies, biological control would not be effective against r pests because of their reproductive potential. The use of insecticides is the best method for the control of r pests. At the opposite end of the spectrum are K pests, which have relatively low fecundity, poor powers of dispersal, specific food preferences, and few natural enemies. They occur in relatively stable habitats and are best controlled by methods which render their habitat less stable, for example, cultural practices and by the breeding of resistant strains of the organisms they attack. Intermediate pests form a continuum between r and K pests and are normally held in check (below the economic injury threshold) by natural enemies. Biological control is the primary method for control of these pests, supplemented as appropriate by chemical, genetic, and/or cultural methods.

Legal control aims to prevent or control pest damage through legislation. It includes the setting up of quarantine stations, systems for monitoring pest populations, and mechanisms for the certification of disease-free plants and animals.

Chemical control traditionally has been the use of naturally occurring or synthetic chemicals to kill pests. It has been the major method of pest control for about 70 years but has created three serious problems:

(1) a great increase in the resistance of pests to the chemicals;

(2) the death of many beneficial insects due to the chemicals' nonspecific activity, and

(3) pollution of the environment.

More recently, increased interest has been shown in insectistatics and sex attractant, which interfere with normal growth and reproduction.

Biological control is the regulation of pest populations by natural enemies. For insect pests, parasitoids, predaceous insects, and microorganisms are the major control agents. Biological control offers several advantages over control by insecticides:

(1) the control agent is ready-made (does not have to be developed), cheap to produce and apply, and persistent;

(2) the method does not stimulate genetic counterattack by pests; and

(3) it is specific, and does not endanger humans or wildlife through pollution of the environment.

Its main disadvantages are its slowness of effect and the fact that the final (equilibrium) pest population density is normally higher than that achieved after insecticide application.

Methods of genetic control include (1) those which render pests less capable of reproduction, for example, the sterile insect release method, use of chemosterilants, and sterility resulting from mating incompatibity or abnormal gametogenesis, and (2) those in which resistance is increased in the organisms attacked by pests.

Cultural control includes the long-established agricultural practices which make habitats less suitable for pests. The methods used may either directly affect a pest, stimulate an increase in density of a pest's natural enemies, or make the organisms on which a pest feeds more tolerant of attack.

Integrated control (pest management) is a combination of methods for reducing and maintaining pest populations below the economic injury threshold. To be most effective it requires the input of as much information as possible about the agroecosystem under consideration and the collaboration of experts from a variety of disciplines. If conducted properly, integrated control programmes lead to considerable financial saving and a great improvement in environmental quality.

6. Literature

The literature dealing with economic entomology is voluminous, and the following list is highly restrictive. Descriptions of economically important inssects (mainly North American species) may be found in standard textbooks of applied entomology for example, Metcalf *et al.* (1962), Pfadt (1971), and Little (1972). Problems associated with the use of insecticides are discussed in the Proceedings of the XV International Congress of Entomology (1977), in volumes of the Annual Review of Entomology, and in the volume edited by Metcalf and McKelvey (1976). Biological control is the subject of the volumes edited by Burges and Hussey (1971), Huffaper (1947), and Huffaker and Messenger (1976), of DeBach's (1974) text, and of reviews by several authors in the Proceedings of the XV Congress. Metcalf and Luckmann (1975), Apple and Smith (1976) and various contributors to the Proceedings of the XV Congress have discussed integrated control (pest management). The genetic control of insects is reviewed by Whitten and Foster (1975). Selected aspects of genetic control are discussed in the Proceedings of the XV Congress.

12

The Fungus Gardens of Insects

Anyone with at least a passing interest in biology is aware that fungi, being plants that lack chlorophyll and cannot conduct photosynthesis, live on other organisms or on decaying organic matter. It is less well known that many insects are similarly dependent on fungi. Indeed, there are insects that tend elaborate gardens of a fungus, controlling the growth of the plant according to their own specialized needs.

Some insect species are always found in association with a certain fungus, and some fungi only with a certain insect. Such complete interdependence is called mutualism. The mutualistic partners of insects are not limited to fungi; they include other microbial forms such as bacteria and protozoa. In some cases the insect feeds on its partner or on the partner's own partly digested food. In others the partner lives in the insect's alimentary tract and digests food the insect cannot digest for itself; frequently the partner supplies an essential constituent that is deficient in the insect's diet, such as nitrogen or a vitamin. Some mutualistic partners serve more than one of these functions.

The insects that are mutualistic with microbial organisms are divided into two groups. In one group the fungus bacterium or protozoon lives inside the insect, either in the alimentary tract or in specialized cells. In the other group a fungus lives in the insect's nest. Here we shall discuss the relations between fungi and insects in the latter group, leading up to those insects that actively cultivate fungus gardens. Such relations have been studied for more than a century, but they offer many new possibilities for investigation. One wants to know more about the physiology of the relations, about how the partners interact at the molecular level. A deeper knowledge of the physiological mechanisms would undoubtedly clarify how the mutualistic partnership came to be established in the course of evolution. It might also have important by-products. For example, much work has been done on the possibility of using fungi that are harmful to insects as a means of selectively controlling insect pests; such work might be advanced by knowing more about the relation between insects and beneficial fungi. As a second example, those insects that control the growth of fungi in gardens may do so by means of antibiotic substances that might well be useful to man.

The first kind of insect-fungus relation we shall take up centres on the tumor-like galls that sometimes appear on the bud, leaf or stem of a plant. These galls develop when certain insects deposit their eggs in the plant and somehow cause it to form an abnormal tissue which then nourishes the larva that emerges from the egg [see "Insects and Plant Galls," by William Hovenitz; Scientific American, November, 1959]. The galls caused by the mosquito-like midges of the family Itonididae also contain fungi. Growing parasitically on the gall tissue, these fungi usually form a thick layer on the inside of the gall. They appear at an early stage of the gall's development, and a single species of fungus is consistently found in association with larvae of each midge species. Many of these fungi, however, also grow independently of the insect. How fungus and insect come to be together in the gall is not known, but some workers believe the female midge deposits spores of fungus at the time she lays her eggs. Many of the fungus galls are caused by insects the feed by sucking plant sap, and except for a few cases it is unlikely that the fungus acts directly as a source of food.

It may be that the fungus assists the insect indirectly by partly breaking down the gall tissue so that the insect can digest it.

Many plants bear insect-fungus galls but only a few of the fungi have been identified. Some galls we have studied in our laboratory at the University of Kansas are leaf-blister galls on several kinds of goldenrod and aster caused by at least nine species of the midge Asteromyia (all of them associated with the fungus Sclerotium asteris in the U.S.), and flower-bud galls of the broom (Cytisus) caused by the midge Asphondylia cytisii (associated with the fungus Diplodia in the U.S. and Europe).

In contrast to the casual association between insects and fungi in plant galls, several species of the fungus Septobasidium and various scale insects (family Coccidae) that inhabit the fungal tissue coexist in a manner that is clearly mutualistic. Septobasidium resembles a thick lichen: it clings tightly to the leaves or branches of trees. The scale insects that inhabit this fungus in some way modify its growth in their vicinity, giving it a different texture or colour; as a result some colonies of Septobasidium have a mottled surface.

The relation between this fungus and the insects that colonize it has been described by John N. Couch of the University of North Carolina. The insect, which feeds on sap, is attached to the tree by its sucking tube. The mycelium of the fungus—its thick mat of fine threads—shelters the insect from the weather and shields it from birds and parasitic wasps. In turn a few of the insects are penetrated by specialized threads, called haustoria, that extract nourishment from the insect's blood. Scale insects characteristically ingest more sap than they need; the fungus may take advantage of this fact by utilizing such nutrients while they are still circulating in the insects's body.

Septobasidium is distributed by scale insects as well as nouiished by them; it does not live independently in nature. When the insects are young, some of them crawl on the surface of the fungus and become covered with spores. A few of the contaminated insects migrate to new areas on the tree, where they insert their sucking tube. Shielded by new mycelium, they survive. They are also invaded, however, by haustoria from the developing spores, and as a result they do not attain maturity and reproduce. The new mutualistic colony is nonetheless able to continue because uncontaminated insects now find shelter in the mycelium. In some species of Septobasidium the entire process has apparently been made more efficient by the development of hollow "insect houses" that attract and hold the migrating scale insects. Thus the fungus furnishes a shelter for the insects and the insects provide both a food supply and a means of dispersal for the fungus. Some insects are sacrificed for the benefit of the colony as a whole; both fungus and insect benefit at the expense of the tree.

A different kind of mutualism has been observed involving on the one hand the wood wasps of the genera Sirex, Tremex and Urocerus and on the other the fungi Stereum and Daedalea. These common fungi sometimes build a semicircular "bracket" out from a tree. The adult female wood wasp deposits her eggs in moist wood by means of a long, slender ovipositor, and at the base of this organ are tiny pouches that contain fungus cells called oidia. When the egg is deposited, oidia cling to it. Then the mycelium of the fungus grows into the wood, and when the wasp larva emerges from the egg it follows the path of the mycelium. The fungus partly digests the wood before it is eaten by the larva. In the laboratory wood wasp larvae have been reared on a diet consisting only of Stereum, but whether the fungus is essential to the insect's nutrition is not known. Both in nature and in the laboratory the fungus grows well without the help of any insect.

The wood wasp nonetheless acts as an agent for the dissemination of Stereum. A larva that later develops into a female has organs that ensure the preservation of the fungus. These organs are tiny pits hidden in folds between the wasp's first and second abdominal segments; in them bits of fungus are trapped in a waxy material. Here an inoculum of fungus remains viable, whereas fungus in wood is inactivated when the wood eventually dries out. When the larva metamorphoses into a pupa, the organs that hold the fungus are discarded. Then, when the adult female wasp emerges from the pupal skin, tiny flakes of fungus-impregnated wax that have fallen from the organs become lodged in the moist pouch at the base of the ovipositor. The flakes now give rise to a fungal mycelium, which develops the oidia that will coat the eggs. The transfer of the fungus from generation to generation of wood wasps is thereby assured.

Ambrosia beetles also carry fungi within their bodies. The numerous species of these wood-boring insects (families Scolytidae, Platypodidae and Lymexylonidae) cannot survive without ambrosia fungi (several genera of Ascomycetes and "imperfect" fungi). In their external skeleton are small pockets called mycangia (literally "fungus containers"); these pockets always contain a supply of viable fungus spores. When an ambrosia beetle tunnels into wood, spores are dislodged from the mycangia, and soon a mass of velvety fungus lines the interior of the tunnel. On this "ambrosia," which concentrates in its cells nutrients that have been extracted from the wood, the beetles feed.

Ambrosia beetles have been the subject of considerable research because they destroy much timber all over the world. Their tunnels, sometimes called "shot holes," extend deep into the sapwood of trees and are surrounded by streaks of a stain manufactured by the enzymatic action of the fungus. The beetles most often attack hardwoods, preferring trees that have been weakened by drought, disease or fire, or fallen timber that is moist and filled with sap. Attracted by the odor of fermenting sap, the beetles fly upwind, usually at dusk; it is easy to collect them in the evening around a newly felled log. They are similarly attracted by the yeasty smell of beer and beer drinkers, and it is also convenient to collect them at a beer picnic! Some kinds bore into beer and wine kegs, which is why in Europe they are called "beer beetles."

The tunnels of ambrosia beetles can be distinguished from those of other wood-boring insects by a black or brown discolouration of the wood around the neat circular tunnel opening. There is, moreover, no wood dust or fecal matter inside the tunnel. When the beetles are excavating, fine wood particles, sometimes mixed with the insect's brown feces, accumulate outside the tunnel entrance. The beetle does not as a rule eat wood, and it rids its nest of wood borings. The males of some species assist the females with tunnel excavation. Inside the ambrosia beetle's tunnel system one finds, depending on the species, either separate niches, each enclosing a single glistening larva or a pearl-like egg, or several larvae sharing an enlargement of the tunnel. In many species the adult insect, on emerging from the pupal skin and proceeding to feed on the mass of ambrosia fungus lining the tunnel, rock's back and forth in a curious manner. What this does is force fungus spores into the mycangia before the insect flies away to found its own nest.

Each species of ambrosia beetle is normally associated with only one species of ambrosia fungus. Ambrosia fungi are pleomorphic: they can readily change when their growth medium is changed, from a fluffy moldlike form to a dense yeastlike form. In the mycangia and the tunnels of the ambrosia beetles the yeastlike form prevails. Recently we have discovered that ambrosia beetles can also change the form of other fungi from the moldlike form to the yeastlike one. This is a significant phenomenon, and we shall be returning to it.

The most conspicuous and most destructive of the fungus-growing insects are termites. The termites that cultivate fungi are native to the Tropics of Africa and Asia. In west Africa it is estimated that the yearly cost of repairing the damage done by these insects to buildings is equal to 10 per cent of the buildings' value. In addition to wood the insects eat growing and harvested crops and objects made of rubber, leather and paper; they destroy documents, works of art, clothing and even underground cables. The enormous mounds that some termite species build for nests interfere with farming and hinder road construction. If they are incompletely destroyed, the insects rebuild them.

Many species in the genera Macrotermes and Odontotermes make their nests in spectacular steeple-like mounds of hardened earth, which in Africa reach a height of as much as 30 feet. Other species, in the genus Microtermes, are completely, subterranean, and if it were not for their mating flights and the damage they do, they would be quite inconspicuous. Each nest contains a white, sausage-like queen and a king, usually enclosed together in a protective cell of earth; the much smaller workers, soldiers and young termites (nymphs) of various ages and both sexes are found throughout the nest. At certain seasons winged male and female reproductives (future kings and queens) are also present. In each nest are one or several fungus gardens, the number and shape depending on the species of termite. Material collected by the workers is chewed and swallowed, and the partly digested fecal material is deposited on a fungus garden when the workers return to the nest. In the nests of some species there is a single large mass of fungus garden; in one of the nests of Odontotermes obesus that we studied in India the mass was two feet in diameter and weighed 80 pounds. In other nests many fungus gardens one or two inches long are scattered along burrows throughout the nest. Each fungus garden is enclosed in a close-fitting cavity lined with a mixture of saliva and dirt. The chambers of some species are ventilated by an elaborate system of vertical conduits extending to the surface of the nest [see "Air-conditioned Termite Nests," by Martin Luscher; Scientific American, July, 1961].

The fungus gardens look like a gray or brownish sponge or may be convoluted like a walnut meat. They are moist but usually firm and brittle, and are permeated by threadlike mycelium. Seattered over the surface and inside the pores of the gardens are numerous minute, glistening, pearly white spherules composed of masses of rounded fungus cells.

The role of the fungi in the nutrition of the fungus-growing termites is not clear. It is well known that many common

Temperate Zone termites cannot digest the cellulose in the wood they eat but rely on certain cellulose-digesting protozoa that live in their intestines to do it for them. No protozoa live in the gut of the fungus-growing termites, therefore it seems likely that the fungi growing in the gardens break down cellulose and may provide vitamins also. In fact, these termites soon die on a diet restricted to wood. The termites continually eat away the fungus gardens as they add fresh fecal material. There is thus a communal interchange of food, the fecal material being partly broken down by fungi in the garden, then eaten again and redeposited in the garden for further digestion by the fungi.

The white spherules are frequently picked up by the workers and moved to other parts of the fungus garden or are sometimes eaten. The king, queen, young nymphs and soldiers are apparently fed saliva by the workers; no trace of fungus or plant material can be found in their digestive tract. Some winged reproductive termites contain material from the fungus gardens, with which they may begin a new garden when they start a new nest.

Several genera of unrelated fungi grow in the fungus gardens of termites, the most abundant are species of the mushroom Termitomyces. The various species of Termitomyces are found only in nests of fungus-growing termites. Somehow the fruiting structure—the mushroom—of the fungi is not allowed to grow from fungus gardens while they are being actively tended by termites. If the termites are removed or die, however, some of the spherules grow into the mushrooms of Termitomyces.

In Africa and southern India termites in many nests simultaneously remove the outer layers of their fungus gardens and spread crumbs of them in a thin layer on the ground during the rainy season. Soon Termitomyces mushrooms appear, and after their spores have been disseminated by the wind the termites come to the surface to collect fungus that may have grown from a mixture of spores from many nests. It is believed that in this way the termites provide for the cross-fertilization of their fungi, as man does for corn and other crops.

Closely resembling the fungus-growing termites in behaviour but only distantly related to them are the fungus-growing ants (tribe Attini). Here we have an example of convergent evolution, in which two nearly unrelated animals and their fungi occupy a very similar ecological niche. These ants are found only in the Western Hemisphere, and most of them are tropical. Some species are found in the deep South and the Southwest of the U.S.; a small species (Trachymyrmex septentrionalis) is found in sandy areas near the Atlantic coast as far north as Long Island.

Atta texana, which lives in eastern Texas and southern Louisiana, does considerable damage to citrus groves, gardens and plantations of young pine trees by cutting leaves from them for its fungus gardens. This ant is known locally as the "town ant," and it inhabits an ant metropolis that is sometimes 50 feet across and 20 feet deep. (One student of these ants, John C. Moser of the U.S. Forest Service, opens their nests with a bulldozer!) Other species of fungus-growing ants build smaller nests; some are so small that they are extremely difficult to find.

The fungus-growing ants probably represent the most advanced stage in the evolution of fungus gardening because they feed only on the fungus, and they actively cultivate it. The workers, depending on the species, collect caterpillar excrement, fallen flower anthers and other soft plant debris as well as leaves cut from trees. Rather than eating the material the ant cuts it into pieces and adds them to a fungus garden in the nest. The ants' gardens look somewhat like those of the termites: they are gray, flocculent masses of finely divided moist plant material loosely held together by threads of mycelium. In the underground chambers of the nest the fungus garden often is suspended from the roots of plants. Scattered over the surface of the older parts of the garden are white specks just visible to the unaided eye. These specks, called kohlrabi bodies, are clusters of bromatia, the swollen tips of the filaments forming the mycelium. The kohlrabi bodies look much like the white spherules found in the gardens of termites. When the ants are not feeding on the bromatia, they lick them.

The flying, nest-founding ant queen carries a small pellet of fungus in a pouch below her mouthparts, mueh as the ambrosia beetle carries fungus in its mycangia. In starting a new nest the young queen grows a small fungus garden on her excreta; with this she feeds the first worker larvae. When the workers are mature, they leave the nest to gather the material with which the garden is enlarged. They feed bits of bromatia to the larvae that nestle in the mycelium of the gardens.

As long as the ants actively tend the garden the fungus does not develop fruiting structures, but mushrooms of four genera have been found growing from abandoned nests of some species or have been cultured from fungus gardens in the laboratory. We do not know how the ants control the growth of their fungi so that they produce bromatia and nothing else; perhaps it is by constantly "pruning" away excess growth of the mycelium. It is also possible that the excreta and the saliva of the ants, which are deposited on the fungus garden, contain some substance that influences the growth of the mutualistic fungus and inhibits the development of the many spores that accidentally enter the nest.

With the fungus-growing ants our brief survey of the mutualistic associations berween insects and fungi ends. It can be seen that there are two distinct kinds of relation. In the gardens of wood wasps, ambrosia beetles, termites, ants and probably those of nudge galls the fungus extracts nourishment from a substrate and the insect feeds either on the fungus, the substrate predigested by the fungus or both. The fungits is prevented from producing sexual frttiting structures but is supplied by an insect partner with a suitable habitat and a means of dispersal. In the colonies of the fungus Septobasidium and scale insects the situation is reversed: the insect feeds on the substrate and nourishes the fungus, and the fungus provides shelter for its castrated insect partner.

Insects are unique among animals in having developed mutualistic relations with fungi. This may have come about because so many insects and fungi share the same tiny habitats. Moreover, most insects are equipped to carry living spores of fungi, either in their gut, in folds between their joints that contain waxy secretions or among their bristles.

The fungus-gardening insects convey into their nest the spores of many fungi other than the one on which they depend. If the insects are removed, the alien fungi will grow and soon overrun the nest; they do not grow in the nest, however, when the insects live there. Apparently the fungus-gardening insects either secrete or excrete antibiotic substances that prevent the growth of alien fungi. The substances may also act to transform the mutualistic fungi, causing either ombrosia, spherules or bromatia to appear rather than sexual fruiting structures.

In the case of the termite, which licks the spherules of its fungus and encloses the fungus garden in saliva-moistened mud, the saliva may contain the substances in question. We have tested the effect of adding saliva taken from termites to a culture of their fungus, which under ordinary growing conditions in the laboratory does not produce spherules. After saliva was added to the culture sphesules grew; the saliva also inhibited the growth of alien fungi. We have performed other tests on the excreta that ants deposit on their nest gardens. Although we have found that the excreta inhibit the growth of certain bacteria, much experimental work remains to be done. The saliva of the ant, which also licks its fungus, may help to form bromatia.

Spores of the ambrosia beetle's fungus remain in the yeast-like form while they are carried by the insect, and it is possible that the waxy secretion might also affect the form of fungi in the tunnel, where the beetle and the fungus are in close contact. On the other hand, it has been shown that the mutualistic fungi of some species of the beetle can be modified to the yeastlike form by certain physical conditions and in the absence of the beetle. These physical conditions duplicate conditions in the beetle's tunnel, where the feeding insect steadily mows the tips of the fungus as they grow.

In nature the presence of a living insect partner is necessary to maintain ambrosia, spherules or bromatia, but these forms can be produced in the laboratory on special mediums in the absence of the insects. When the mutualistic fungi are grown on ordinary carbohydrate-rich laboratory mediums, fluffy mycelium and sometimes sexual fruiting bodies appear. If the same fungi are grown on acid mediums that are rich in amino acids, and are exposed to more than .5 per cent carbon dioxide, then bromatia, spherules or ambrosia are formed. These cultural conditions apparently resemble conditions in the nests of the insects. The fungi of some ambrosia beetles also become ambrosial if they are repeatedly scraped or are grown at low temperatures. Clearly the problem of how insects control the growth of a fungus partner remains an intriguing one.

13

Water Resources

Human use of land depends on the availability of water, and the lack of water is a limiting factor in many, many regions. Where water resources are relatively abundant, this constraint will seldom be confronted, and then only during extended periods of drought. Where water is not so abundant, this constraint is confronted all the time, so the necessity for water management is recognized by most people in such regions. One should not conclude from this that water-resource management is only a concern in dry areas. It is not. Although water-resource management is such an integral part of the functioning of any society that it may be taken for granted, it is always a matter of professional concern.

The Hydrologic Cycle

Every examination of water resources begins with an examination of the hydrologic cycle within the region of interest. The general features of this cycle are represented pictorially which shows the pathways followed by water initially falling on the surface as precipitation. Some of this precipitation returns to the atmosphere through evaporation and plant transpiration. This part of the cycle is referred to as evapotranspiration. Some of the precipitation also moves overland as runoff to streams, lakes, and in some areas, the sea. And finally, some of the precipitation moves down into the ground through the zone of aeration, where the soil is not saturated with water. This infiltrating water may reach the level where the soil or rock is saturated with water. This is known as the zone of saturation, the upper surface of which is called the water table. Water reaching this level becomes part of the ground water. Each of these elements is linked to the others. A surface-water body such as a stream is fed by direct precipitation, by surface runoff, and by ground-water flow. The atmosphere is fed by evaporation from the ground surface, rivers, lakes, and the ocean, and by plant transpiration as well.

A major task of the hydrologist is to measure the flow of water from one point in the hydrologic cycle to another. These data must be gathered over time since these flows all vary with time. The analysis and use of hydrologic data showing variation with time are central to hydrology. Water flow is viewed in the framework of a hydrologic system consisting of many separate elements linked to one another by pathways along which water moves. A diagrammatic representation shows more clearly than the pictorial representation the different pathways that water can follow to arrive at any one destination.

Hydrologic information comes from studies by meteorologists and hydrologists. In the United States, the collection and recording of these data are primarily the responsibility of the U.S. Geological Survey Water Resources Branch and the U.S. National Weather Service, although many government agencies and other institutions also help. These data are published in USGS water-supply papers, hydrologic atlases, and other series prepared by the Water Resources Branch. The USGS also publishes the National Water Summary, which identifies hydrologic events and related issues in all 50 states (U.S. Geological Survey, 1984).

We will first consider how these measurements are made. Then we will examine some general information on water use. Next we will look at issues of surface-water management.

Finally, we will look at issues of ground-water use. Although this division is convenient, there are many instances in which we must consider the relationship between ground water and surface water because the two are closely connected.

Measurements Employed in Hydrology

Here we consider some aspects of measurement and the units that are employed by hydrologists in reporting these measurements.

We can think of the flow of water from one element in the hydrologic system to another as if it took place in a pipe connecting the two elements. Water enters this "pipe" from one element, moves through the pipe at some rate, and is discharged from the pipe into the next element. A simple example is represented by two ponds connected by a river. The river represents the pipe in which water is transported from the upper pond to the lower. The discharge into the lower pond is measured in terms of the volume of water released in a given time. The discharge from one end of a pipe may be expressed in units of L^3/t, where L is a unit of length and t is a unit of time. Common units used to express discharge are cubic feet per second (ft^3/sec), generally written as cfs, and cubic metres per second, generally written as m^3/sec. In the United States, hydrologists also use an additional unit, the acre-foot. An acre-foot is the volume of water that would result if an area of one acre was to be covered by 1 foot of water.

In general, instantaneous discharge is expressed as the product of the cross-sectional area (A) of the "pipe" and the mean velocity of flow of water through the "pipe" at the point of discharge. Measuring instantaneous discharge requires measuring the cross-sectional area and the average velocity of flow.

It is typical relationships in streams in regions that are not arid. Note that the cross-sectional area of the water-filled channel increases downstream. Because of this, and because stream velocity generally increases downstream, the volume of water passing an upstream point at a particular moment usually is less than the volume of water passing any point further downstream at the same moment. The figure also shows how the areas of the water-filled cross-sections increase during periods of increased precipitation.

Measurements of discharge are made at locations called gaging stations, where the cross-sectional area of the water-filled part of the channel can be measured directly. At such a station, the mean velocity of flow of water past the station may be determined by measuring the velocity at several different points and depths and then averaging these by appropriate procedures. The cross-sectional area is designated as A and the mean velocity as v.

The instantaneous discharge, designated by the letter symbol Q, is then the product of the measured values of A and v at the time of measurement:

$$Q = Av \tag{13.1}$$

It is apparent that the cross-sectional area of the water-filled channel increases as the water level in the river rises, and it decreases as the water level falls. Thus rising water level leads to increasing discharge; and falling water level leads to decreasing discharge. As the water level rises, the mean velocity of flow also increases, which contributes to increased discharge.

A record of the fluctuation of discharge with time is called a discharge hydrograph. The U.S. Geological Survey publishes hydrographic information of this kind for thousands of rivers in the United States in its water supply paper series. The discharge values at each gaging station are measured by continuously monitoring the water level or stage of the river. The measured value of stage is converted to a value of discharge by the use of calibration curves, called stage-discharge or rating curves, which are based on a study of the relationship between the cross-sectional area (A) and stage, and between the mean velocity (v) and stage, at each gaging station. Once these calibration curves have been prepared, it is no longer necessary to measure cross-sectional area or stream velocity. The simple measurement of stage is sufficient.

Another important measurement determines the volume of water in storage in a reservoir or in a segment of a stream channel. It is easy to picture the water in storage in a reservoir. If we know the configuration of the bottom of the reservoir, we can determine the volume of water in that reservoir for any observed water level. The same concept applies to a river channel, even though water is moving through that channel, and it is perfectly proper to speak of water in storage in a channel.

Finally, we are often interested in changes in storage during time intervals of observation, especially in monitoring water supplies in reservoirs. These changes in storage in a reservoir. If we know the configuration of the bottom of the reservoir, ($\triangle$S) per unit time is equal to the difference between the average rate of inflow (I) to the reservoir and the sum of the rate of outflow (O) and average evaporation (E) during that time. For finite periods of time, the equation is as follows, where $\triangle t$ is the time period of interest:

$$I\triangle t=(O\triangle t+E\triangle t)+\triangle S \qquad ...(13.2)$$

With this basic terminology in hand, we can examine the subjects of water use and surface-water management.

Water Use

Every region is characterized by its own natural hydrologic cycle, which existed before the region was settled or developed. Humans occupying that region use water moving through various elements of the local hydrologic system for agriculture, livestock, industry, power development, natural resource development, and human consumption.

Much of the water drawn from one element of a local hydrologic system is not returned to it. For example, much of the water withdrawn for irrigation is lost to the atmosphere through evapotranspiration. Such a use is called a consumptive use. Many other uses allow almost all of the water to be returned to the same element from which it was withdrawn. For example, river water diverted through an electrical power generating plant is passed through the power-plant turbines and returned to the river. Such a use is called a nonconsumptive use. Many other uses allow for the return of part of the water withdrawn. For each such partially consumptive use, we can describe the fraction lost in terms of net consumptive use. These terms and their application are explained fully in USGS Circular 1001 (U.S. Geological Survey, 1983a).

The water returned to an element of the surface-water system helps to maintain the volume of water in that element, However, this returned water is likely to be warmer than, or chemically different from, the water that was withdrawn. As a result, this returned water is likely to change the water quality in that element. Ordinarily, such changes in water quality are seen as adverse, and are therefore described as causing a decline in water quality.

The distribution of water among users in the United States may be seen from a diagram the distribution and changes in distribution of water use in the United States. In the United States in 1980, the average rate of water withdrawal was 450 bgd (billion gallons per day). Of this total, 150 bgd went to irrigation and 255 bgd to industry. Most of the water used by industry was for cooling thermo-electric power plants (210 bgd). Of all the water withdrawn each day, about two-thirds were returned to surface-water bodies. The other third was returned to the atmosphere as water vapour or was incorporated in manufactured products. Irrigation was the dominant consumptive use of water, accounting for 83 bgd in 1980. The sources of the 450 bgd were surface water (64 per cent), ground water (20 per cent), and saline waters (16 per cent).

From our point of view, these national patterns of water use are of less interest than patterns of water use within a single area or region. The reason for this is simple—in most regions, the water used comes from supplies within that region. Where this is true, issues of water-resource management are initially local or regional issues, not national issues. Thus, if we want to learn how scientific information is employed in dealing with water-resource issues, we need to learn about local water-use patterns and supply problems.

In a particular region, water-use patterns evolve by accident, design, or some combination of accident and design. And once a pattern is created, it is not easily broken. As a consequence, we seek to ensure that the existing water-use pattern can be maintained and even expanded. To do this usually requires an intricate interplay between scientific knowledge, technological manipulation, and a body of water law and custom. Where water-resource problems arise, they will be dealt with, if at all, through the use of scientific information tightly constrained by the legal system and political considerations. We can see this by examining the hydrologic features of the Colorado River basin and the water use in that basin. This provides a case history and an introduction to basic concepts that must be dealt with in any examination of surface-water management.

Case History: Surface-water Management in the Colorado River Basin

Geography and History

The Colorado River basin is a 609,520 km² area, within which surface runoff and ground water move toward the Colorado River and its tributaries. The basin is divided by law (the Colorado River Compact of 1922 into upper and lower basins. The entire basin lies within an arid region. Average annual precipitation in the upper basin is about 16 inches and in the lower basin is about 8 inches. Most of this precipitation is lost to the atmosphere through evapotranspiration. In the upper basin these losses are estimated to be 85 per cent and in the lower basin even more than this. Yet the upper basin supplies 83 per cent of all the water that flows down the Colorado River. The upper basin supply is obviously of critical importance to all users of Colorado River water.

Serious consideration of the use of scientific information in water-use management in the United States began with the work of John Wesley Powell in the Colorado River basin. In 1869, Powell explored the Colorado from the mouth of the Green River in Wyoming to the Gulf of California. Later he prepared a major study of the arid regions of the United States (Powell, 1878), in which he presented his case for water-use management in these regions. As the second Director of the U.S. Geological Survey, he made strenuous efforts to convince the U.S. Congress of the need to create a scientic data base for the Colorado River Basin.

Powell's scientific work is important because it led him and others to the view that land-use development of the region should take such factors as water availability into account. His personal history (Rabbitt, 1969) is of interest as well, because it portrays the conflicts that can arise between a scientist and his opponents. These conflicts are obvious in all efforts to manage the Colorado River.

Basic Hydrologic Information for the Colorado River

The hydrologic system of the Colorado River Basin functioned in its natural state before the basin was occupied by European settlers. The flow in the Colorado River under such natural conditions is referred to as the virgin flow, and the term "virgin flow" is used today to describe the flow that would occur in a river if there were no human influences.

All efforts to manage water withdrawals and to design systems of dams depend on the availability of a record of natural flow—also called virgin flow—in the river, a record that will show the short- and long-term variations in flow caused by natural phenomena. From this hydrographic record we can obtain such important additional information as the annual high and low flows, the total annual flow in each year, and the mean annual flow over the period of record. For the lower Colorado River, such information is obtained at Lees Ferry, Arizona, where a USGS gaging station is located.

Flow in the Colorado River has been monitored at Lees Ferry since 1921, and flow at some upstream points has been monitored even longer. This record is one of the best bases on which to formulate predictions of the range in scale of future flows in the Colorado River.

The flow record the measured annual flow at Lees Ferry, called actual flow, and the estimated virgin annual flow. The actual flow is the total volume of water measured in millions of acre-feet (maf) passing through the river in a 12-month period. If no water had been wlthdrawn upstream, then this actual flow would be the same as the virgin flow. Since upriver diversions have ranged between 2 and 4 maf each year, the observed flow has generally been 2 to 4 maf less than the estimated virgin flow.

Some of the important hydrologic facts that bear upon water-management decisions. We see first of all that the virgin annual flow during the period of record has varied strikingly from a low of 5 maf to a high of 24 maf. The overall mean annual virgin flow computed from the entire record is 14.9 maf. However, if the record is divided into two parts, we find that the mean annual virgin flow at Lees Ferry for the 35-year period 1896-1930 was approximately 17 maf; and for the 35-year period 1931-1965, it was approximately 13 maf. This basic record leads us to ask some questions:

1. Has there been a long-term decrease in the virgin flow of the Colorado River?
2. Are the estimates for the earlier 35-year period less reliable than the estimates for the more recent period? If they are, then we cannot conclude that there has been a long-term decrease.

3. Do years of low flow or high flow tend to occur in groups.

All of these questions have been addressed repeatedly by analysts (see National Research Council, 1968, 1977, 1981) because the answers are of critical importance in assessing past and future plans for water management.

The scientific information considered in the preceding paragraphs was obtained by direct measurement during a relatively short period of historic record. Efforts have been made to expand this record and to put it into perspective by using tree-ring chronologies for a 400-year period back to 1564 A.D. (Stockton, 1977). Studies of tree rings allow estimates to be made of precipitation in the basin; and these estimates serve, in turn, as the basis for estimating virgin flow in the Colorado River. A sample of this record suggests that there have been periods of tens of years in which the virgin flow was well below 13 maf, and that there have been individual years in which the virgin flow was below the historic low flow recorded in 1934.

This information, which seems to be so basic, takes on considerable importance when we examine the pattern of past use and the plans for future use of water from the Colorado River Basin.

Water Withdrawals and Water Management Problems

The water-management problem, simply stated, is to determine how water supply and water use can be managed so that existing patterns of water use are served, and where new patterns are proposed, to find new schemes of accommodating them.

The water-management problem began to take shape in the Colorado River Basin during the years in which Powell was active as a U.S. Geological Survey scientist. Development of the problem is reviewed in many sources. Here the principal sources are the USGS Professional Papers 441 (Iorns, Hembree, and Oakland, 1965) and 486 (Hely, 1969; McDonald and Loeltz, 1976), the National Research Council studies 1968,

1977, 1981), the California Water Atlas (Kahrl, 1978) and the historical review of Rabbitt (1969).

The problem was created by the successive diversions of water from the river for consumptive uses, which were largely agricultural; this has resulted in progressive diminution of river flow. The first such diversions took place in 1840 in the northern part of the basin, when a Mormon settlement in Wyoming diverted water from the Green River, a major tributary to the Colorado, to provide water for irrigation. By 1880, the Uncompaghre Valley had been put under irrigation; and in 1909, the U.S. Bureau of Reclamation constructed the first diversion tunnel, a 6-mile long tunnel carrying water from the Gunnison River to the Uncompaghre.

Diversions in the southern part of the basin were started at about the same time as those in the north, and they soon came to dominate the pattern of consumptive use. Therefore the focus here is on the lower Colorado, the segment down-river from Lees Ferry, Arizona, in the southern part of the basin. The first diversions in the southern part began in 1877 when Colorado River water was diverted to the Palo Verde Valley in California. In the 1890s, the Yuma Valley in Arizona was placed under irrigation. Then, in 1901, the Imperial Valley, in California, which had been called the Imperial Desert, began to receive water fed from the Colorado via the Alamo channel, which had been cut through Mexico for a distance of 80 km.

Development of the Imperial Valley represented the first truely large-scale diversion from the Colorado River. In 1903 and 1904, the Alamo channel was blocked by deposits of silt. A new channel was made, and the channel cut became enlarged in 1905 during a period of floods on the river. This new channel became large enough to allow all of the Colorado River water to flow into the Salton Sink, which thereby became the Salton Sea.

Early diversions for agricultural use were later augmented by new diversions for metropolitan water-supply purposes.

These were begun in 1941 with completion of the Colorado River aqueduct, which carried water Lake Havasu, created by the Parker Dam, to Los Angeles. By 1961, this aqueduct was serving the water-use needs of 7,500,000 people. These and the many other diversions are, which shows the dams below Lees Ferry and the major diversions.

Additional diversions are planned under the authority of the Colorado River Basin Project Act of 1968; which included authorization for the Central Arizona Project. This project is intended to supply 1.2 maf of water to the Phoenix and Tucson areas when it is completed. These diversions may seriously deplete flow in the river. The reservoirs created to store water for diversion will also increase evaporation, further depleting flow.

The Framework of Colorado River Basin Water Law

The water-use management system for the Colorado River has evolved under the influence of a set of legal controls created by the U.S. Congress and the Supreme Court and under the influence of the basic principles of water-resource law in the western United States. The key principle in western water-resource law is represented by the doctrine of prior appropriation, which says that a user who has once established a pattern of beneficial use of previously unappropriated surface water has, in so doing, established a future right to continued water use on the scale initially established. Thus, the early use of Colorado River water in the Imperial Valley represents the establishment of a pattern of beneficial use of previously unappropriated water.

The legal controls established by the U.S. Congress and the Supreme Court constitute a preface to our examination of the application of scientific findings about the Colorado. The more important controls include:

1. The Colorado River Compact of 1922,

2. the Boulder Canyon Project of 1928,
3. the Rio Grande, Colorado, and Tijuana Treaty of 1944,
4. the Decree of the U.S. Supreme Court, 1964, Re: Arizona v. California, Colorado River, and
5. the Colorado River Basin Project Act of 1968.

These are reviewed in Hely (1969), Kahrl (1978), and the National Research Council (1981). Here we need only sample them to see the problems ahead.

The Colorado River Compact of 1922 apportioned water in the Colorado River system to an "upper basin" and a "lower basin", and it also provided for the satisfaction of any water rights of Mexico that the United States might recognize at any future time. One of the most important provisions of the compact is that the upper and lower basins are each allowed to allocate 7.5 million acre-feet of water per year to consumptive use. Another provision is that the upper basin must ensure that 75 million acre feet of water are allowed to flow past the measuring point (called the compact point) as Lees Ferry, Arizona, during any 10-year period. This provision reflects the knowledge that most of the flow in the lower Colorado River is derived from the upper Colorado River. The compact did not allocate water among competing users.

The Rio Grande, Colorado, and Tijuana Treaty awarded 1.5 maf to Mexico, placing an additional demand on Colorado River water. Allocation of water among competing users is represented by the decree of the U.S. Supreme Court of 9 March 1964, which specified that "of such 7,500,000 acre-feet of consumptive use, there shall be apportioned 2,800,000 acre-feet for use in Arizona, 4,400,000 acre-feet for use in California, and 300,000 acre-feet for use in Nevada." If less than 7,500,000 acre-feet are available in a given year, then it is the responsibility of the U.S. Department of Interior to apportion the use among the three states. This provision may be put to a test when the Central Arizona/ Project is completed since the 1968 legislation authorizing this project requires that California first be assured of receiving 4.4 maf per year before any water is diverted into the Central Arizona Project system.

We see from these provisions that water is allocated by law. These allocations reflect existing patterns of development and use. Can these allocations be met in each and every year in the future? If they cannot, then major economic and political problems will be faced in those years in which the allocations cannot be met. The natural and social scientists who have studied the Colorado River system have the task of assembling factual information about the present system and developing information about the history of flow in the system. Another goal is to predict the future behaviour of the system.

Scientific Information and Water Management

The Colorado River, like most large rivers, is a managed river. It is managed to provide water for irregation, power generation, municipal use, and recreation. We have seen that the first stages in the evolution of this water-use management system were taken before there was much scientific information available. Even in 1922, when the first legal controls were being placed on the system, decisions were made to allocate water in the belief that at least 14.8 maf were available for distribution in an average water year. With this compact as a framework, a water-management system was developed.

One of the major features of the water-management system is a series of dams that created nine reservoirs along the Colorado. These reservoirs have a total capacity of 61.6. maf. However, no more than 56.4 maf are supposed to be stored in them, so that 5.2 maf of additional storage capacity will be available for use when there is a threat of flooding. With this system, water can be stored during periods of high flow and then can be released from the control points. These released are intended to provide for the diversions prescribed by law. The reservoirs can provide the water needed to meet the legal requirements during periods of low flow. This same system is also designed to provide for electrical power generation and flood control in the lower Colorado. Inevitably, these functions can come into conflict. If the reservoirs are kept filled to meet water demands during dry periods, they are unable to provide needed flood-control storage during periods of high flow. This situation was faced in 1983, when severe flooding was experienced in the lower part of the Colorado River Basin. Scientific management of surface-water resources is concerned with finding the best possible compromise solutions to meeting conflicting demands such as these (see National Research Council, 1982).

Unfortunately, it now appears that the system may not be able to supply all the water needed for uses that now exist, not to mention those that are being planned for the future. In other words, scientific realities for the Calorado may place constraints on the exercise of legally assured rights and on land and resource-use developments being considered.

The major part of virgin flow in the Colorado comes from the upper basin, and the future of the entire Colorado River Basin therefore depends on the magnitude of that flow, on withdrawals within the upper basin, and on the legal requirement that the upper basin deliver at least 7.5 maf per year to the lower basin. Let us look at these figures.

In 1974, water consumption in the upper basin was estimated to be 3.7 maf, divided among various uses. The Bureau of Reclamation has estimated that at least 5.8 maf should be available for use within the upper basin in any year. The Colorado River Compact actually specifies that 7.5 maf are to be available for use in the upper basin by legal right.

However, the record of virgin flow at Lees Ferry suggests reasons for concern. For example, if average annual flow at Less Ferry in the next 25 years is only 13.1 maf, which was the average annual flow from 1931 to 1974, then the maximum amount of water available for consumptive use in the upper basin would only be 4.9 maf per year (13.1 maf—8.25

maf=4.9 maf). The figure of 8.25 maf is used by Dracup (1977) instead of the 7.5 maf to allow for commitments in addition to those in the 1922 Compact. Use of the 7.5 maf figure would, of course, indicate that more water would be available for consumptive use in the upper basin. If average flow during a period of 5 to 10 years were to be only 10 maf per year, as it has been in the past, then the maximum amount of water available to the upper basin would only be 1.75 maf per year (10 maf—8.25 maf=1.75 maf). This is about half of the volume actually used in 1974.

Clearly, in an unmanaged river, there is no way in which the existing commitments and current consumptive uses could be met in the event of a period of low flow. As we have seen, the Colorado is managed by control of flow at the various dams. Let us look at the way in which this managed system might function during a period of relative drought, following Dracup's example.

The major dam that provides significant control in the upper basin is the Glen Canyon dam just above Lees Ferry, which is designed to provide a usable storage capacity of 25 maf in Lake Powell. The total upper-basin storage capacity is 31 maf. This stored water is intended for use during periods of relatively low flow.

Suppose that total flow in a 10-year period was 100 maf, providing an average annual flow of 10 maf. During this 10-year period, the upper basin must release between 75 and 82.5 maf (Dracup, 1977, p. 131) to the lower basin, leaving at a minimum only 17.5 maf available to the upper basin during this 10-year period. In addition, let us suppose that Lake Powell is full at the start of this drought. In this event, the total water available to the upper basin in the 10-year period will be 48.5 maf (17.5 maf+31 maf—48.5 maf). This amount would allow the upper-basin use patterns of 1974 to be met; however, it would not allow the planned use patterns of 1995 to be met.

These estimates are based on the optimistic view that Lake Powell would be full at the onset of drought. However, if Lake Powell is not filled to capacity, then less water will be availabie during an extended drought. For example, in 1974, upper-basin storage was only 23.6 maf. In the event of a 10-year drought that begins with this level of storage, the net water available for use in the upper basin would be only 41 maf (17.5 maf+23.6 maf), or 4.1 maf per year for 10 years. This is even less adequate to meet future needs.

Calculations of this sort, carried out using existing records or synthesized records of possible future flows, allow the water managers to estimate the probability that a given water commitment cannot be met in any future year or period of years. If such calculations show that the probability of such a deficit is increasing, then more attention can be given to finding alternative solutions to water-supply problems.

Alternatives for Manipulating Water Supply and Water Use

Even though the Colorado River water-supply system is being placed under stress, all analysts agree that there are alternatives available to cope with that stress. Part of the task of these analysts is to provide the best possible estimates of the probability that supply constraints will be felt. They must also estimate the probable duration of a period of such constraints. Another part of their task is to identify and describe alternatives. Some of the alternatives they have examined include:

1. Importation of water into the Colorado River system from such sources as the Columbia and Snake rivers in the northwestern United States.

2. Changing the existing patterns of land use so that strongly consumptive uses such as agriculture are replaced by less-consumptive uses.

3. Decreasing evaporation losses from canal and reservoir systems.
4. Decreasing evapotranspiration losses by eliminating phreatophytes—water-loving plants of low economic value.
5. Terminating any further withdrawals from the Colorado such as the withdrawals planned for the Central Arizona Project.
6. Terminating plans for energy-resource development in the upper basin.

These and other alternatives have been presented for analysis by the committees of the National Research Council (National Research Council, 1968, 1977, 1981), and some of them have been examined rather carefully. We cannot consider every alternative, but we can consider some to illustrate the kinds of questions that are raised.

The northern states, including northern California, are not yet as close to the limits to be confronted in a drought period as are the states of the southwestern United States, including southern California. Therefore proposals have been made repeatedly to export water from the northern states to the southwestern states. A model for such transfers has been created within the state of California, which has already developed a system that will allow it to transfer water from the north using the California Aqueduct System. Proposals to pipe water across state lines to the Colorado have met firm opposition. This opposition is recorded in the 1968 legislation authorizing the Central Arizona Project, which contains a provision explicitly forbidding federal studies of diversions from north to south.

Water demand in the Colorado River basin could be kept in check by changing the patterns of land use. For example, agriculture in Arizona consumes approximately 90 per cent of the water used in the state. Conversion of agricultural land to other uses is already beginning to decrease water use on this land. Such changes can be brought about by changing the fee structure for Colorado River water, which has allowed agricultural users to purchase Colorado River water for much less than metropolitan users. In addition, the cost of ground water use is rising because the water table is dropping. Small farmers cannot afford to pump from deep wells, and they may abandon farming as a result. Changes might also be brought about through the sale of water rights held by agricultural users. Such a sale might take place if energy-resource development in the upper part of the Colorado River basin promises a greater economic return than agriculture, or if such energy development becomes a critical national need. However, this kind of shift would cause a decrease in water use only if energy development used less water than agriculture.

Those who believe that the Colorado River cannot supply all existing commitments during dry periods also argue that the Central Arizona Project (CAP) should be terminated to prevent the planned diversion of 1.2 maf/year from the Colorado River to central Arizona. Analysts have argued that the $2 billion capital costs of this project cannot be justified if water will not be available to feed the system. However, in 1981, construction was begun on this project in spite of these concerns.

Finally, the alternatives of cycling used water back into the system and of augmenting the system with ground water may be considered. Water used for irrigation is partially recycled into the system after use. Unfortunately, this water is usually more saline than the water withdrawn from the river. For example, water taken from the Colorado at Yuma, Arizona, had a salinity of 800 parts per million (ppm) in 1961, but water returned to the river had a salinity of 6000 ppm. Since a 1973 agreement between the United States and Mexico limits the salinity of Colorado River water delivered to Mexico, such recycling faces a severve limit. This has already led to the construction of a desalinization plant near Yuma, which will desalinate water returned to the Colorado from the Wellton-Mohawk agricultural district.

Similarly, augmenting the system by drawing upon ground water has already been found to be a limited solution for two reasons:

1. where ground water is itself relatively saline, its use only adds to the burden of salt in the river; and
2. the ground-water supply in Arizona is actually being mined—an expression used to indicate that ground water has been taken from supplies beneath Pheonix and Tucson at a rate exceeding any possible rate of replacement of recharge.

As a result, the water table has been dropping for years, creating the specter of unavailability of the very water upon which development of the region was predicated. It was just this decline in ground-water that led to the proposal for the Central Arizona Project.

These changes are graphically demonstrated on maps of the entire region that show the decline in water levels and the decreases in the saturated thickness of the aquifer. The history of this decline is shown in a plot of water-level elevation against time.

These declines have already had significant consequences. The cost of raising water from ever-deeper wells has increased steadily, and as a result, there has been a steady decline in the acreage under irrigation. Landowners with water rights to the ground-water resource have been compensated by the federal government for the decline in this resource through a provision in the U.S. tax laws called the depletion allowance. This provision allows water consumers to decrease the taxes they must pay in proportion to the loss of the water resource over which they have control.

Since there is no other water resource in the region that can support the high historic levels of agricultural use, the region can be expected to go through substantial changes in land use and productivity. The Geological Survey has made some preliminary estimates of the rate at which water levels can be expected to decline, given various levels of irrigation use. These estimates, or contingent predictions, can be employed in economic analyses examining the future of agriculture in this region.

These predictions have led to discussions of the future possibility of importing surface water from the Great Lakes or the Mississippi River. Thus recognition that the ground-water resource is declining is having the same effect that we saw for the lower Colorado River.

Water Law and Water Policy

In the development of water resources of the Colorado River Basin, government policies and judicial decisions played an extremely important part in determining the allocation of water. Although these decisions were made by the Congress and by federal courts, there is no uniform body of law at the federal level that governs water use and water rights. Most laws governing water use are made at the state level, and this creates a complex situation nationally. Let us look at broad principles and then at some details of law at the state level.

Two basic approaches to water law are followed in the United States. One of these is based on the concept of riparian rights; the other is based on the concept of prior appropriation. The concept of riparian rights dominates eastern water law; the concept of prior appropriation dominates much of western water law.

Riparian rights are based on the English common-law view that water is a common-property resource that may be tapped by those who control land along the river bank or lake shore. The same principle applies to ground-water resources; the owner of land overlying an aquifer also owns the right to tap that aquifer and withdraw water from it.

The doctrine of prior appropriation takes a different point of view, grounded in Spanish law and practice. Under this doctrine, the right to withdraw and use water is based on historical usage. Someone who puts a certain amount of water to beneficial use establishes the future right to use a similar amount of water. Water is treated as real property, and property rights grow out of prior use or appropriation. Under this system, the initial user is empowered to sell the established water rights to a future user.

These two doctrines grew out of patterns of use of surface-water resources, but they have been employed in establishing rights to ground-water resources. Severe problems have arisen in many, many regions through the exercise of these rights, so individual states and regions within states have created bodies of law to deal with these problems or have tried to get users to collaborate with each other in managing the resources. We will sample the development of water law and policy in several regions.

The most sharply drawn questions concerning the need for water law arise in arid regions where ground water is being mined. Arizona is one such area; Texas and other High Plains states make up another such area. The dominant water law in these regions is based on the concept of prior appropriation. This doctrine has not limited ground-water use because each land owner has been free to extract as much water as was technically and economically feasible. This practice has created problems for contiguous land owners, since drawing down the potentiometric surface or water table affects a subsurface area much larger than that underlying the property owned by the user responsible for the drawdown.

Although this has been known for some time, neither Arizona nor other arid-land states have yet established much water law governing ground-water use. By looking at the development of water law in Arizona, we will see that the original situation is changing. We will also see that water law has a direct effect on the use of land in regions with significant water law.

Until 1980, Arizona had no system for controlling the use of ground water. In June 1980, the Governor of Arizona signed into law a bill controlling water withdrawal and land use. The Governor, Mr. Bruce Babbitt, made the following observation:

> In the old West, we're going overnight from a laissez-faire system, a system where everybody used whatever they wanted wherever they wanted, to the most comprehensive ground-water management system of any state in the American West.

The bill was introduced and passed because the federal government had informed the state of Arizona that it could not draw from the Colorado River until it had taken action to control its use of water. The new law contained the following provisions:

1. Registration of all water wells is required.
2. Water conservation in major urban and agricultural areas is required.
3. A Director of Water Resources is empowered to set per capita consumption limits for cities and to purchase and retire water rights held by farmers, effective in 2007.
4. Development is prohibited where the developer cannot assure water availability for the next 100 years, and the introduction of agriculture in problem ground-water areas is also prohibited.

This kind of regulation was required because prior appropriation and riparian rights had not been adequate to deal with disputes among water users.

Other states have also recognized this problem. California has been the site of major controversies and court cases arising from disputes over water rights (Kahrl, 1978). California has dealt with these disputes in two ways. First, it instituted a system of correlative rights, which made the watershed or the aquifer system the basic planning unit within which water use was to be coordinated. Those within the watershed are assigned rights to allocate water derived from the watershed or the aquifer. At the same time, California developed large-scale water-resource planning, which requires that water be diverted from some regions of the state for use in other regions. This system gives the state the authority to limit the local water use within these regions.

Trelease (1977) has observed that water law in the United States has changed and continues to change as new demands are placed on the law and on water resources. Improved scientific understanding continually shows how the withdrawals from surface and ground-water resources have effects spread throughout a larger system, effects that sooner or later must be taken into account in water law and planning. Thus we can expect that there will be a continuing tendency like that signalled in Arizona to develop systems of water-resource control, systems that will demand more and more information for their design.

14

Contamination of Water

The Geological Survey has studied the spread of contamination from an ordinary municipal landfill located in the Town of Babylon in Suffolk County, New York. This landfill was placed in use in 1947, and since that time it has been used for the disposal of domestic wastes and incinerator ash and some manufacturing waste. The site is located on top of glacial deposits consisting of sand and gravel that are approximately 30-m thick. These deposits constitute part of the surface aquifer. The water table in this aquifer intersects the lower part of the landfill during at least a part of each year. The water table slopes SSE from an elevation of nearly 50 ft above sea level at the landfill to sea level at Great South Bay. Water-table elevation is highest in the late spring each year, and it reaches a low in early winter.

Contamination from this landfill was monitored in an effort to determine the three-dimensional shape of the body of contaminated water, the rate at which ground water is being contaminated, and the difference between the observed contamination pattern and the pattern that would be predicted by ground-water flow models. The monitoring was accomplished by sampling water from 90 wells, each tapping the aquifer at a known depth. The chemical composition of samples from each well was compared with the chemical composition of ground-water from wells believed to tap uncontaminated water. This provided a measure of contamination and a means of mapping the leachate plume. This leachate plume is simply a body of contaminated water analogous with the smoky plumes one sees downwind from smokestacks emitting particulate matter into the atmosphere.

The contaminated water originating in the landfill contains ions dissolved directly from the material in the landfill and also contains compounds produced by biological decomposition of materials in the landfill. Ions monitored by the USGS at Babylon included bicarbonate, sulphate, and chloride anions, as well as sodium, potassium, calcium, and magnesium cations. Each of these ions is also present in normal ground-water in the region, but each occurs in higher concentrations in the leachate. Let us look at results for a single anion – the chloride ion.

Chloride is present in high concentrations in water leaving the landfill—at least as high as 1300 mg/l (milligrams/litre). It is useful as a tracer ion because it does not interact with the sediments through which it is passing, nor does it react with other substances in the ground-water. Such an ion is described as a conservative ion. A map of the chloride concentration in samples from wells tapping intermediate depths (60 to 80 ft below the water table) shows the shape of the plume, the variation in chloride concentration in water within the plume, and the orientation of the plume. Samples taken closer to the surface are not contaminated, showing that the plume has a cylindrical form. The lower boundary of the plume is located at the boundary between the glacial aquifer and the underlying Gardiners Clay, a layer through which the leachate cannot move.

These relationships were interpreted in the following way by Kimmel and Braids. First, precipitation moves down through the landfill and leaches contaminants from it. Then the leachate moves down into the glacial aquifer. This leachate does not flow in a steady stream. Instead, it forms "pockets" of contaminated water that are denser than the surrounding ground water. At these pockets descend to a level affected by the regional ground-water flow system, they move within the regional ground-water flow pattern. The average velocity of flow of the ground-water may be computed from Equation:

$$v = K(l/n) \times (\Delta h/\Delta L)$$

where v=average ground-water velocity, K=hydraulic conductivity, $\Delta h/\Delta L$=hydraulic gradient, and n=porosity. For the Babylon landfill, these values are: K=470 ft/day, n=0.25, and $\Delta h/\Delta L=0.0021$. The average ground-water velocity calculated from this equation is 4 ft/day.

As the leachate moves in the regional flow field, it mixes with the ground water to form what becomes the plume. The distance between the leading edge of the plume and the source increases with time. As the leachate moves through the aquifer, the concentration of the tracer decreases due to the phenomenon of dispersion. This decrease in concentration in the downstream direction is readily evident from the concentration contours for chloride. It is evident from this figure that there has been very little lateral or transverse dispersion, since the plume width is about the same as the width of the source landfill.

Note that the apparent velocity of leachate movement at Babylon, as measured from the length of the leachate plume, is much less (1 ft/day) than the estimated ground-water velocity. Kimmel and Braids discuss this discrepancy and note that it may arise because of a delay between the time a landfill is created and the time the leachate enters the ground-water flow field. This discrepancy indicates that caution must be exercised in predicting the exact time of arrival of leachate at any point. In spite of this uncertainty, it is clear that the plume will continue to extend itself to the southeast. It may feed contaminated water into Santapogue Creek, or it may flow under the creek and discharge only into Great South Bay. Even after the landfill ceases operation, it will continue to serve as a source of contamination for many years. Covering the landfill with impermeable material would cut off inflow of the leachate. But material that had entered the aquifer up to the time of covering would move slowly downstream, so that many tens of years would elapse before the ground-water could return to normal.

This example shows how a fairly large-scale monitoring operation of a simple system can provide an improved understanding of the process of point-source contamination. It also provides all the information needed to design a model for predicting the contamination pattern in the vicinity of a landfill source. Let us look at an actual case in which the possible uses of prediction have been demonstrated.

Case history: the Rocky Mountain arsenal disposal site. The Babylon case shows how a leachate plume can be monitored by using a conservative ion such as chloride. No strongly toxic materials were present, and the site was monitored to provide scientific understanding of the process. Where toxic substances are known to be present, monitoring a nontoxic conservative ion may help in mapping a target area in which to investigate possible contamination by the toxic substances. If the toxic substance is not conservative, then the contaminated volume of ground-water should be no larger than the volume of water in the plume mapped using the conservative ion. This may be seen from a consideration of ground-water contamination at the Rocky Mountain Arsenal in Colorado as reported in studies by Walker (1961), Konikow (1977), and Konikow and Thompson (1984).

The Rocky Mountain Arsenal, north of the city of Denver, Colorado, is a centre for the manufacture of materials that are designed for use in chemical warfare. Between 1943 and 1956, liquid wastes resulting from the manufacture of these materials were discharged into unlined disposal ponds constructed on an alluvial surface aquifer. These wastes consisted of inorganic and organic chemicals. After they had been mixed with one another in the disposal ponds, they reacted chemically to produce toxic materials such as 2,4,5—T, materials that are believed not to have been present in the wastes as discharged. The wastes were also rich in chloride.

The diposal ponds are located at elevations between 5100 and 5200 ft above sea level. At the time the wastes were placed in these ponds, it was known that the water table in the surface aquifer sloped northwest from the arsenal to the South Platte River. The area between the arsenal and the South Platte River was used for grazing and for crops. Water to support these uses was drawn from wells tapping the surface aquifer and from an unlined irrigation canal. Enough information was available when the pits were first put in use to have predicted that any contamination entering from the pits would migrate toward the South Platte River. No efforts to make such predictions have been reported.

Wastes were first placed in the pits in 1943, and it appears that they began to contaminate the surface aquifer shortly after (Walker, 1961). However, it was not until 1951 that reports of contamination of the irrigated area began to appear. Beginning in 1951, damage began to appear in crops irrigated by water drawn from the surface aquifer. The worst damage was reported in 1954, a particularly dry year. Additional reports of damage to crops and livestock appeared in the Denver Post in 1973 and 1974. Also new contaminants were reported from wells sampled at that time; these included disopropylmethylphosphonate (DIMP), a nerve-gas by-product, and endrin, aldrin, and dieldrin.

All the major studies of contamination summarized by Konikow (1977) demonstrated that the Rocky Mountain Arsenal was the source of ground-water contamination affecting a large area between the arsenal and the South Platte River. Konikow (1977, and earlier reports cited therein) set out to provide accurate maps showing the extent of contaminated areas. He also developed a model for predicting the spread of contamination from the disposal site and for predicting the decrease in contamination that might result from employing various mitigation tactics.

He showed that he could use his model to predict the areal extent of contamination at any specified time. This may be seen by comparing a map of predicted contamination with a map of the observed contamination in 1961. This modelling showed that ground-water flow models could have been used to predict contamination downstream from the disposal ponds.

The model that Konikow developed was based on the use of a conservative anion and could be expected to show the maximum size of the area contaminated. The predictive map could be used as a guide for locating monitoring sites for non-conservative contaminants. Also similar models could help in deciding where to place waste disposal sites of any kind. However, more detailed modelling of non-conservative substances would require that sorption and reaction phenomena be taken into account.

Contamination of Surface Waters from Multiple Sources

Surface waters can be contaminated from the whole range of sources listed earlier in this chapter. Mine workings and mine tailings constitute major point and non-point sources of contamination. These commonly cause acidification of surface waters. This results from the oxidation of sulphide minerals and the resultant formation of sulphuric acid. The contamination is similar to that described for areas affected by acid precipitation. Excellent sources on acid-mine drainage are found in the National Research Council's report on coal mining.

Sewage treatment plants and industrial outfalls are sources of contamination that have caused many serious problems in the past. The most common of these problems is the depletion of dissolved oxygen in the receiving waters, a depletion caused by the anaerobic decomposition of the introduced wastes. Less common, but quite serious, problems have also resulted from the release of toxic chemicals such as Kepone (James River,

Virginia), PCBs (Hudson River, New York), dioxin (Times Beach, Missouri), and mineral particles such as asbestos (Lake Superior). Many of these sources of contamination have been brought under control. As these controls have been introduced, it has been recognized that other, less easily controlled sources of contamination are contributing to water-quality problems. These new sources are atmospheric, as we have seen for acid rain, and terrestrial.

The effect of multiple or dispersed sources is well illustrated by the problems of lakes, bays, and estuaries in areas subject to human influence. These bodies of water are subjected to intense use and are affected by loading from many contaminant sources. The problems are obvious in the many small lakes that the U.S. Geological Survey has described as urban lakes and real-estate lakes in the series on Water in the Urban Environment. Similar and larger-scale problems are also recognized in lakes as large as the Great Lakes, in bays such as Chesapeake Bay, and in estuaries. It is convenient to examine small lakes here since they illustrate processes that act on many scales.

Any lake may be viewed as a complex ecosystem consisting of a physical framework comprising lake basin, soil, and water, and a biota of plants and animals. A continuous exchange of chemical elements and compounds takes place among all parts of the ecosystem as shown diagrammatically. This exchange is facilitated by mixing of upper lake waters (epilimnion) and deeper waters (hypolimnion) in the fall and spring when the epilimnion is cooled or heated, causing changes in the density of the surface water. (The terms epilimnion and hypolimnion are used to describe the stratification of lake waters that results from temperature and density differences.) When these fall or spring changes occur, the lake waters "turn over" and become mixed.

Under natural conditions, the plant and animal assemblages in a lake display a considerable degree of stability; changes in the character of the lake do take place, but they occur very slowly as seen from a human perspective. Each lake evolves through stages that end with the filling of the lake basin. This evolution ordinarily lasts thousands of years.

In the early stages, lake waters are rich in oxygen, and the concentrations of dissolved solids are relatively low. In such lakes the productivity of plants is limited by the availability of nutrients, and this in turn places limits on the size of animal populations. These lakes are described as oligotrophic. As lakes evolve in the natural state, the waters become more nutrient-rich, and a more abundant plant life develops. At first this increased plant life makes possible a more abundant and varied assemblage of animals as well. As the production of phytoplankton and algae increases, however, a stage may be reached at which the dissolved oxygen content of the lake waters is reduced. This changes the character of the lake drastically. We say that a lake has become eutrophic when this stage has been reached.

This evolutionary process may be represented diagrammatically in terms of the production of organic matter per unit area of lake surface. During the oligotrophic stage, the rate of production of organic matter increases slowly. Then, during the natural eutrophic stage, the rate of production increases rapidly. Finally the rate of production decreases, and the lake fills with sediment and becomes extinct.

Where human influence is significant, the eutrophication process may be accelerated, increasing the rate of production of organic matter very rapidly. As a result, a lake can quickly enter a state that would take thousands of years to reach under natural conditions. Let us examine this process.

Lakes and bays that become surrounded by urban development show a phenomenon called nutrient loading, created by inputs of nutrients from point sources such as sewage treatment plants (STPs), and by inputs of nutrient-rich runoff from agricultural and residential land on which fertilizers have been applied. As a result of this loading, lake waters may become so rich in nutrients that the production of algae is no longer limited by the availability of one or more critical nutrients. When this happens, phytoplankton productivity may cause algal blooms, which cover the lake waters with mats of algae. Algal material in these mats dies and sinks below the surface, decomposing as it sinks and consuming oxygen from the subsurface waters. Then the decomposition products are released to the surface waters. This process drastically changes the character of the lake. Fish kills may occur, and the fish population may also change through a modification of the kinds of species present and through a decrease in the total species variety.

Many urban lakes and reservoirs are also affected by being filled rapidly with sediment eroded from the watershed. The basins of small lakes can be completely filled in periods measured in years (Rickert and Spicker, 1971). In addition, the incoming sediments carry adsorbed nutrients and a whole array of contaminants. Once these sediments come to rest, they constitute a source from which nutrients and contaminants can be released over time This is important because it may affect efforts to restore lakes that have been partially eutrophied and contaminated, as described in the next section on restoration.

Somewhat comparable changes are observed in much larger bodies of water, where such changes take on greater economic importance. This is illustrated by changes taking place in Chesapeake Bay, which is a large estuary fed by a number of rivers draining the eastern seaboard of the United States. These rivers include the Susquehanna and the Potomac, both major rivers draining heavily urbanized watersheds.

Chesapeake Bay is an important commercial fishery and recreational resource. Research showing long-term decreases in oxygen content of bay waters and changes in plant productivity have stimulated much concern. Officer *et al.* (1984) have shown that the volume of Chesapeake Bay waters that becomes anoxic (oxygen-free) in the summer has been increasing rather steadily since the 1950s, if not earlier. This may be seen from maps of

Chesapeake Bay for the summers of 1950 and 1980, which show that by the summer of 1980, most of the Chesapeake Bay bottom was anoxic (oxygen-free) of hypoxic (very low dissolved oxygen).

Officer and co-workers argue that these chemical changes in Chesapeake Bay waters apparently are having large-scale ecological effects, causing declines in commercially important benthic organisms such as oysters, and changes in the populations of organisms on which commercial fish species feed. They attribute the anoxia to increased nutrient input into the bay.

The changes in the composition of lake waters and in the character of lake biota are difficult to reverse. Because of this, it is widely agreed that the best way to deal with such problems is through watershed management that controls input before trouble can arise. However, many lakes have already been accelerated to a eutrophic stage. Where this is true, efforts of restoration may be considered as explained in the next section.

Control and Restoration

In the preceding sections, we have seen that there are many, many overlapping processes that change ground and surface waters adversely. Possibilities for limiting these processes from the start have been recognized, and restoration procedures have been identified. We will consider these possibilities and procedures next.

Each region, when confronted with contamination of soil, ground water, rivers, lakes, estuaries, and coastal waters, must employ whatever tactics appear to be technically, economically, and politically feasible. These tactics may be directed at preventing future contamination, restoring already contaminated systems, restricting use of contaminated areas, and compensating those affected adversely. Let us first examine control measures and then consider restoration measures.

Control of Point-Source Contamination

Contamination of ground water takes place at countless dumps, landfills, disposal pits and ponds, and injection wells. Contamination of surface waters takes place wherever an outfall pipe discharges contaminants into rivers. lakes, and estuaries. These sources are point sources that can be controlled. We will look at some of the control measures that can be taken in the sitting of such facilities.

Sanitary landfills and various kinds of disposal pits are the most common sources of ground-water contamination. Siting of new landfills or pits is therefore a subject of intense interest in thousands of communities. The Babylon, Long Island, landfill is an example of a sanitary landfill; and the pits at Rocky Mountain Arsenal and the Love Canal disposal sites are examples of toxic-waste disposal sites. An examination of sanitary-landfill siting and design will serve as a guide to the general principle to be considered for all of these facilities.

A sanitary landfill is a site at which domestic and municipal wastes are placed so that they do not create a public nuisance and do not create hazards to public health. In practice, wastes are trucked to the site, dumped onto the ground surface in a trench or against an embankment, and covered with some kind of earth materials. As the lankfill grows, it typically takes the form of a truncated pyramid rising above the surroundings. Within this pyramid are cells consising of refuse, each more or less completely separated from other cells by the earth materials used as cover.

One of the best introductions to the technical details of landfill siting and hydrogeology is in the work of the Illinois Geological Survey. One of their reports (Hughes, Landon, and Farvolden, 1971) is a major source for the following material.

The potential for contamination from landfills is determined by the kinds of wastes contained in them and by decomposition processes acting on the wastes. Decomposition is accomplished by microorganisms that act under aerobic conditions in the uppermost part of the landfill and under anaerobic conditions in the lower parts. Carbon dioxide and methane are produced by this decomposition, and the methane migrating upwards is an important contaminant. A liquid, called leachate, forms from the refuse and contains high concentrations of solids and organic materials.

When precipitation infiltrates a landfill, the leachate is carried downwards to the water table. This is an intermittent process. If the base of the landfill intersects the water table, then the leachate can enter the ground-water continuously. Therefore the goal in the design of landfills and disposal sites is to maintain release of the leachate or liquids at an acceptable level and to place the site so that escaping contaminants will be attenuated or will follow paths that avoid contaminating the surrounding areas. Hughes, Landon, and Farvolden have summarized the tactics needed for achieving these goals:

1. Eliminate the production of leachate.
2. Allow the leachate to migrate within an acceptable area.
3. Provide for recovery of the leachate at a distance from the landfill.
4. Retain the leachate so that it can be recovered.

Let us examine how each of these tactics may be implemented.

Elimination of Leachate Production. If the infiltration of precipitation into the landfill can be controlled, leachate production may be partially controlled as well. The most economical way to achieve this is to cover the upper surface with the least permeable materials available. Such materials may be glacial till, clay-rich materials, or even man-made cover such a asphalt. Such measures do not totally prevent the formation of leachate, since ground water may enter the landfill even if no water infiltrates from above. Therefore it is desirable to situate landfills at sites above the water table. Such sites are common in arid regions, but they are scarce in humid regions.

Migration of Leachate under Acceptable Conditions. Since most landfills produce and release leachate, it is important to situate them so that the leachate will move slowly and become decontaminated as it moves. The Illinois studies (Hughes, Landon, and Farvolden, 1971) have shown that clays, glacial tills, and unfractured shales make excellent landfill substrates. Ground-water moves very slowly through these materials, and total dissolved solids are often reduced by a factor of 10 to 100 as the leachate travels a distance of a few metres.

Where materials having these properties are not to be found, then other approaches must be considered. At sites like that at the Babylon landfill, where the materials are highly permeable sands, one option is to place the site so that the flow paths terminate at a body of surface water where dilution may result. At Babylon, the flow paths may pass beneath Santapogue Creek and terminate at Great South Bay. Another option is to place the site so that flow paths guide the leachate into a deep aquifer containing water that is known to be of unacceptable quality to begin with.

Migration and Recovery. Where local flow systems are well understood, it may be feasible to locate a landfill so that the leachate will be carried along flow paths converging at a discharge point where the leachate can be captured as needed. For example, a landfill placed on a hill above flow lines terminating in a local swamp or pond might contaminate that swamp or pond without affecting the larger regional groundwater system. If monitoring along the flow path showed that contamination levels were reaching unacceptable levels, leachate collection wells could capture the leachate for treatment.

Retention and Recovery. A landfill or pit can be designed so that a leachate recovery system is in place from the beginning of use. Such a system might consist of burried perforated pipes, called tiles, laid horizontally, or wells placed in or around the site. These perforated pipes and wells could be pumped to withdraw the leachate. As they are pumped they would also alter the local flow system, preventing the leachate from entering the larger ground-water system. This tactic is illustrated in a cross section that shows the flow system under natural conditions and the flow system that would result from pumping from a burried tile system.

Restoration of Contaminated Systems

Throughout the world there are cases in which soil and water have been contaminated. New cases are reported daily in the United States and presumably in other industrialized countries. As each case arises, there may be a need for dealing with the problems faced. In many instances, the need is acute, and quick solutions to the problem must be found. At Love Canal, the quick solution was evacuation followed by compensation. But as the number of cases grows, larger-scale solutions must be found. Some of these will require the development of programs to restore and rehabilitate soils, aquifers, and surface waters. Others call for the development of programes of compensation. In most instances, these programes demand a legislative and financial base that is created regionally or nationally. Here we will consider different kinds of restoration and rehabilitation. Policies that make these programes possible will be considered in the concluding sections of the chapter.

Examples of Ground-water Restoration. Once a contaminant has spread through a large volume of an aquifer, that contaminant can be removed or diluted only slowly, given the slow rate of ground-water flow in most systems. Yet some efforts have been undertaken and others surely will be.

At Love Canal, New York, the New York State Department of Environmental Conservation employed one of the tactics described in the preceding sections. Trenches were dug around the disposal site, and drainage tiles were placed in these. Contaminated ground water moving into these tiles was pumped from them, and this water was then fed to a water-purification system-in which activated charcoal was used to remove most of the contaminants from the water. The purified water was then fed to the Niagara Falls municipal treatment system. This system is described in detail in the U.S. Environmental Protection Agency Report (1982).

The Environmental Protection Agency has subsequently monitored wells around Love Canal for the various organic chemicals that had been found there earlier. The EPA found that the drainage-tile water-purification systems had apparently contained the spread of contamination. The initial cost of this system was large ($ 8 million), and the costs of continuing to operate it are substantial. Thus, although this tactic can work, it is expensive. Furthermore, the toxic chemicals captured on charcoal must also be disposed of safely to avoid simply transferring the problem to a new site.

The system at Love Canal was installed after a large-scale evacuation. A similar approach is being followed at another Niagara Falls disposal site, but under somewhat different conditions. The site is the Hyde Park site north of Love Canal, a site also operated by the Hooker Chemical Corporation. After a court case brought by the U.S. Government against Hooker, an agreement was made that stipulates in detail the nature of the monitoring system that is to be placed around the site, the leachate collection system that is to be installed, and the conditions under which this system is to be operated. This legal document illustrates the use of hydrologic information in the courts (see United States of America v. Hooker Caemical and Plastics Corporation . . . Civil Action No. 79-989, Stipulation and Judgement Approving Settlement Agreement, U.S. District Court for Western New York, 19 January 1981).

At Rocky Mountain Arsenal, where the source of contamination consisted of liquid wastes in open pits, a different combination of tactics has been employed. Here, the first step was the construction of new pits having impermeable liners. The waste was pumped from the older pits into these new pits.

This stopped the further infiltration of contaminated water. Then the older pits were filled with fresh water and kept filled to provide the hydraulic head needed to drive fresh water through the surface aquifer. Eventually this fresh water should dilute and displace contaminants from the aquifer. In addition, burried walls were constructed to impede the flow of ground water. Wells were placed on the upstream side of these walls to capture contaminated water. This water was treated, and purified water was then recharged through wells on the other side of the walls. The predicted effects of this approach have been described by Konikow. The observed effects have been described by Konikow and Thompson (1984).

These examples show that, to a degree, ground water can be restored. This requires considerable information about ground-water flow and substantial financial support. Restoring contaminated soil and ground-water is much more costly than preventing contamination.

Examples of Lake and Stream Restoration. Substantial efforts are underway to restore the water quality of lakes and streams in the United States. For many rivers the first goal has been to ensure that dissolved oxygen levels are raised to former levels and are maintained there. A second goal has been to deal with contamination by substances such as polychlorinated biphenyls (PCBs) or dioxin. For many lakes the initial goal has been to slow the process of induced eutrophication by controlling algal productivity. An example of each of these efforts is presented here.

The Willamete River in Oregon is one of the most carefully studied rivers in the United States. Each of a continuing series of U.S. Geological Survey Circulars 715A-K concerns various aspects of river quality assessment of the Willamette River Basin (see Rickert *et al.*, 1980, for the most recent in this series). As explained by Rickert and collaborators, "The maintenance of high dissolved oxygen concentration has been the critical problem in the Willamette River" (Rickert *et al.*, 1980, p. Kl). Until rather recently the dissolved oxygen content of river waters was very low during low-flow periods, and low or zero levels of dissolved oxygen hindered the normal migration of salmon upriver.

This is no longer true, and normal salmon migration patterns have been partially restored. This restoration has occurred because a programme of secondary waste-water treatment of all point-source discharges was instituted. The Willamette is the largest river in the United States to be so fully controlled. The USGS has developed modelling techniques that allow the prediction of dissolved oxygen levels at various flow rates and to study the effects of the augmentation of flow or the further control of industrial discharges. This modelling has shown that summer flows must be augmented in order to assure meeting present standards for dissolved oxygen. Any further improvements will depend on the reduction of industrial waste inputs of ammonia in the stretch between river mile 50 (above the mouth) and river mile 134.

The improvement in water quality seen for the Willamette has been seen in many other rivers as well, among them the Hudson River in New York. Improvements of this sort have restored the fish populations for which these rivers were once so well known. In the process, these improvements have revealed the importance of other contaminants, as may be seen from a brief examination of the Hudson River.

The Hudson river has been contaminated by PCBs discharged from plants of the General Electric Company, which used the PCBs in the manufacture of transformers and capacitors. The use and discharge of PCBs are now forbidden by U.S. federal law. Unfortunately, PCBs are long-lasting and continue to enter the food chain. Studies of water and sediment have shown that the PCBs are held by the sediment particles. Therefore two issues must be dealt with in efforts at restoration. The first issue has been to eliminate the discharge of BCBs from the manufacturing plants. The second issue has been to determine how PCBs are moved and buried in river sediments. If the PCB-bearing sediments are covered by uncontaminated sediments and will remain covered, then they may pose little threat and might be ignored. But if flood discharges are expected to periodically erode the sediments and move them downstream to new locations, then the costs and benefits associated with dredging and removal of sediments from the river should be considered. The first prerequisite is to obtain good information about sediment concentrations and sediment movement partners. Preliminary findings of research on the Mohawk and Hudson have been published, but final resolution lies in the future.

The possibility for restoration of lakes and the complexity of restoration are particularly well illustrated by efforts to bring about improvements in water quality of Irondequoit Bay in Rochester, New York. This bay was undergoing cultural eutrophication, probably increased sediment loading, and remarkable loading by sodium chloride during the period from 1950 through the 1960s. As a result of these changes, it was a target for scientific research on eutrophication, salinity change, and restoration.

The restoration measures have taken the following forms. In 1973, New York state banned the use of phosphate in household detergents. Then a large-scale system for phasing out all sewage treatment plants discharging effluent into this bay was planned and constructed. Finally, the use of de-icing salts was reduced within the watershed of the bay. These control measures have brought about decreases in the loading of nutrients and in the chloride-ion concentration in bay waters (Banister and Bubeck, 1978). In spite of these reductions, additional sources of contamination have been identified. One source is the lake-bottom sediments themselves.

Sediments entering in past decades have trapped nutrients on the bottom of the lake. Will the lake sediments release nutrients and other contaminants to lake waters as the input of contaminants from external sources is decreased? The answer is not yet known; research leading up to this is reviewed in Bannister and Bubeck (1978). According to studies done by the

U.S. Geological Survey under the National Urban Runoff Programme (NURP), water quality in the bay has not improved as much as expected. This programme is investigating the contamination contributed by surface runoff and ground-water within the watershed.

Radioactive Wastes—Planning for the Future

Prediction of future risk has not been an important element in the selection of sites for the disposal of most nonradioactive wastes. Thus there are few examples of scientific prediction in the records of site selection and design for the hundreds of sites identified by the EPA as sources of concern.

In contrast with this record, planning for the disposal of the radioactive wastes created by commercial nuclear-fission reactors is characterized by very large-scale, long-term efforts to employ scientific analysis and prediction. This is true in all countries in North America and Europe that have nuclear-reactor programmes, and all of these countries are investing substantial scientific resources in the design of their programmes for radioactive-waste management. For this reason, the scientific basis for radioactive-waste management is worth special examination.

What is radioactive waste? What is it accorded such unusual attention? To answer the first question requires a familiarity with the fission-reactor approach to the generation of electricity. To answer the second question requires a familiarity with the growth of knowledge about the human-health effects of ionizing radiation. Each of these topics is treated briefly here before examining the actual process of scientific analysis of the problem of nuclear-waste disposal.

As the radioactivity from most fission products declines, the radioactivity from the actinides increases; and after a few hundred years, most of the radiation output comes from americium, neptunium, plutonium, and two important long-lived fission products, technetium 99 and iodine 129. The radioactivity produced by each of these elements may also. These elements present a source of radiation hazard for very long periods of time.

One way to appreciate the degree of health hazard caused by these various elements is to examine a diagram that shows the volume of water that would be required at any given time if one were to mix a specific nuclide with sufficient water to create a solution that could be ingested safely. For example, that the amount of strontium 90 present in a reactor assembly containing one megagram (10^6 grams) of heavy metals (uranium, plutonium, etc.) would have to be diluted in more than 10^{10} cubic metres of water to meet the health standard. The fission-product curves that the volume of water required to dilute most of the fission products to safe levels declines to zero after periods of 10 to 1000 years have elapsed. In other words, most fission products are not a source of hazard after periods ranging from 10 to 1000 years.

Radioactive Waste from Reprocessing. The fuel assemblies described above are the form that reactor waste takes if a once-through cycle is employed. These assemblies may be reproces. sed by dissolving them in nitric acid and extracting uranium 235 and plutonium 239 for future use. This extraction leaves most of the fission products and the actinides in the nitric acid solution. The extraction process also frees such gaseous or potentially gaseous elements as carbon, hydrogen, krypton, and iodine from the fuel elements. These must be captured by appropriate techniques, or they will escape to the atmosphere.

Reprocessing creates a collection of radioactive wastes including an acid solution, here referred to as high-level waste (HLW), and separated masses of iodine, hydrogen 3 (tritium), carbon, and krypton. All of these wastes must be dealt with, but since they are in different forms, special waste-handling techniques must be employed. Perhaps the most significant technique is that for handling the nitric acid solution.

To date, commercial wastes from reprocessing have been stored in steel tanks at West Valley, New York. These wastes and others like them must be solidified. This solidification process was begun at West Valley in 1984. The first step in the solidification process is the production of granular material through a process called calcining. Then this granular, highly radioactive material is converted to a stable solid. Two forms may be noted here. The granular material might be subjected to fusion to produce metre-long cylinders of borosilicate glass. Alternatively, it might be heated with appropriate silicate materials to produce centimetre-long rods of SYNROC, a synthetic analogue of natural silicate rocks (Ringwood, 1982). If the fusion technique is employed, the radionuclides are dispersed in solution in the borosilicate glass. If the SYNROC technique is employed, the radionuclides take up specific positions in the crystal structures of the synthetic minerals of which the SYNROC is made. In either case, a stable, reasonably strong, but highly radioactive cylinder is produced that is more easily managed than the nitric acid solution from which it was made.

Geologic Isolation Systems for Nuclear Waste

Basic Goals. Whatever the form taken by the radioactive wastes, they must be isolated from contact with the biosphere. All of the countries included in the Organization of Economic Cooperation and Development (OECD) are at wark on developing systems for the isolation of radioactive wastes in rock bodies beneath the earth's surface. Progress in each of these countries is reported in an OECD study (Nuclear Energy Agency, 1982). Those who are studying these isolation systems must evaluate the problems to be solved in dealing with storage of fuel elements, borosilicate glass, and SYNROC, because it is possible that all three waste forms will be employed. The waste will be placed below ground in a permanent repository developed in rock or sediment. This method of below-ground storage is called the geologic approach to radioactive-waste disposal. It has been adopted in the belief that it provides the best guarantee that radioactive wastes will be kept from contact with the biosphere over periods of thousands of years.

The decisions to be made concerning geologic disposal are land and water decisions of the first rank because they have the potential for affecting land and water use for thousands of years. Some of the criteria to be considered are:

1. The waste package should remain stable for at least 1000 years.
2. The waste package should be isolated from ground water as long as possible.
3. The waste package should be designed so that there will be the least possible exchange with ground water, if contact is made with ground water.
4. The ground-water path between the waste package and the biosphere should be such that radionuclides released from the waste package will not reach the biosphere for at least 1000 years after release.

Scientists and engineers are confronted with unprecedented prediction problems. They are being asked, for example, to provide scientifically based predictions concerning the path that released radionuclides will follow and the geochemical behaviour that these nuclides will display while moving along that path. All such predictions will be "best-possible" predictions rather than certain predictions, so all efforts to design isolation systems include plans to provide multiple barriers to the ultimate dispersal of radionuclides. We will illustrate this by an example.

We will examine the actual state of investigations into geologic disposal systems. This will illustrate both the scientific prediction problems and the issues that must be faced. Here we will look at progress in two countries, the United States and Sweden. Both countries use the same types of reactors, and both have national laws specifying that safe storage systems must be developed to hold reactor wastes. The Swedish laws were passed in 1977; the U.S. laws were passed in 1982 (see below). We will consider Sweden first because it appears to be closer to actually instituting a long-term isolation system.

Swedish Geologic Disposal Systems. In 1977, the Swedish nuclear power industry commissioned Swedish analysts to produce 120 technical reports describing all facets of the radioactive-waste storage problem in Sweden. These reports are referred to as KBS-1, KBS-2, etc. They provided the technical foundation on which two plans were built, one a plan (KBS-I) to dispose of the wastes from reprocessing, the other (KBS-II) a plan to dispose of spent fuel elements. The second plan (KBS-II) was reviewed by a committee of the U.S. National Resource Council, so an excellent summary of the plan is available in English (National Research Council, 1980).

The KBS-II plan proposed that spent fuel elements from all of Sweden's reactors be stored initially at a single central site (centrallager), built in bedrock at a depth of less than 100 metres. The fuel elements would be stored in water-cooled tanks at this site for at least 40 years. Then the assemblies would be placed in thickwalled copper canisters filled with molten lead. These copper- and lead-jacketed assemblies are the form in which the reactor waste is to be placed in the repository.

The final repository is to consist of rooms and tunnels carved out of granite at a depth of 500 metres below the surface. Each copper cylinder will be placed in a hole drilled into the floor of the room. The copper cylinders are to be surrounded by bentonite (smectite) clay. Once the repository has been filled with cylinders, it is to be observed for an indefinite period, during which time it will theoretically be possible to retrieve the cylinders if problems are identified. Finally, the entire repository is to be filled with a sand-bentonite-ferrous phosphate mixture. Once this has been done, it will be impossible to retrieve the cylinders.

The first phase of the Swedish system is already nearing completion. A central storage facility for spent fuel has been constructed in bedrock at the Oskarshamn nuclear power station on the Baltic Sea. Loading of this facility between 1985 and 1990 will provide valuable experience for Sweden and valuable information for the United States, since the United States may need a facility for the temporary storage of spent fuel until a deep repository can be constructed.

The design of the waste package and the final repository illustrates the multiple-barrier system. The copper-lead jacket is expected to remain stable for 1000 years or more. It constitutes the first barrier to the release of radionuclides. The bentonite clay is expected to retard the access of water to the cylinders and to capture any nuclides released. It constitutes the second barrier. The sand-bentonite-ferrous phosphate mixture is also expected to serve as a chemical barrier. It constitutes barrier number three. The design of these barriers is the concern of engineers. Geologists must carry out geological studies to predict ground-water flow in the vicinity of the repository and to assess the general geologic stability of the repository site.

Additional information needed is designing the final repository in Sweden is being obtained from an international research project in progress at the abandoned mines at Stripa, west of Stockholm. This project is designed to provide essential information on ground-water flow in granitic rocks (Witherspoon, Cook, and Gale, 1981). The research done at Stripa has shown that monitoring systems must be installed within the repository itself, since monitoring based only on holes drilled from the surface has been shown to be inadequate.

Geologic Disposal Systems in United States. In the United States, target dates and procedures have been set by the Nuclear Waste Policy Act of 1982 (PL-97-425). This act specifies that the selection, design, and operation of the repositories are a federal responsibility. The act also specifies that the costs of operation of the repositories are to be paid for from a Nuclear Waste Fund created by taxes on the commercial nuclear power industry. The first goal of this policy is to provide storage for spent fuel. By the year 2000, the United States expects 178,000 canisters containing spent fuel to be on hand awaiting disposal. Each canister will be a 0.3 m × 4.6 m package containing either a single PWR assembly or two BWR assemblies. This inventory is expected to completely fill a single deep respiratory.

The Nuclear Waste Policy Act specifies that the President of the United States must recommend the first repository site by 31 March 1987 and the second site by 31 March 1990. The first date by which a construction permit might be appproved is 1 January 1989. Draft criteria have been prepared by the Environmental Protection Agency specifying standards to be met by such a repository. The repository is to be placed more than 500 metres below the surface at the site chosen. Shafts will be sunk, and tunnels will be extended from these shafts.

The geology of the United States is far more varied than that of most of the other OECD countries, and many different geologic and geographic settings present themselves for review. Each such setting must be studied carefully during the initial screening process in order to determine how well it is likely to meet the proposed criteria.

The U.S. Geological Survey is responsible for identifying appropriate settings in which to place repositories. The following kinds of rocks are potentially suitable: salt formations (bedded salt or salt domes), crystalline rocks (granite and gneiss), basalt, rhyolite tuff, clay-rich formations, and unsaturated valley-fill sediments.

These kinds of rocks are present in the United States. In the early years of the commercial nuclear power industry, committees of the National Research Council (NRC) concentrated on salt deposits as one of the most promising types of geologic media in which to place a repository. In the 1980s, the USGS began examining sites that could provide a much greater range of options.

We can illustrate the general approach by considering a setting in which bedded salt deposits are present. Salt deposits have been well studied, and work to date illustrates some of the procedures to be followed. General reviews of salt and all other media are to be found in the geologic isolation study.

The early NRC review committees found salt to be a promising repository medium because salt deposits are generally impermeable to ground water, as may be seen from a visit to a working salt mine. Such a mine is perfectly dry, unlike most other mines. In addition, salt has the highest thermal conductivity of any rock being considered. This is advantageous because the high thermal conductivity ensures that the temperature adjacent to the waste package will be a lower value than it would be in a rock having lower conductivity. Initially, salt was also viwed as a satisfactory medium because it is plastic and tends to flow around or into openings. This crates a self-sealing character to a salt deposit, which is desired. However, it also creates a less stable setting than may be desirable.

Two types of salt deposits have been considered. One consists of layers of sodium chloride interbedded with shales and other sedimentary rocks. This type of deposit is referred to as bedded salt. The other type consists of cylindrical bodies of sodium chloride that are arranged so that the long axis of the cylinder is vertical. These are called salt-dome deposits, because the cylinder has pushed up the overlying sediment layers into domes. Both types of deposits have been mined extensively, so there is a great deal of geologic information available about them.

In the early days of repository investigation, a bedded salt deposit at Lyons, Kansas, was chosen for trials. However, investigation showed that this deposit had been penetrated by too many drill holes to make it suitable for further development, and the Lyons project was abandoned in 1971. A bedded salt deposit at Carlsbad, New Mexico, was then chosen for investigation; it became the site for the first large-scale repository pilot project.

The project at Carlsbad was named the Waste Isolation Pilot Project (WIPP). Shafts have been drilled from the surface down to a level 650 metres below the surface, and a repository is being created in salt layers at that level. This repository is intended only for transuranic wastes created during the production of nuclear-weapons materials, and it is not to become a high-level waste or spent-fuel repository. However, it may be used as a site at which experimental studies of the effects of single canisters of spent fuel or high-level wastes are made.

Studies made at WIPP showed that there are drawbacks to the use of salt deposits that had not been fully recognized earlier. During exploratory drilling in the area surrounding the WIPP, large pockets of brine (highly saline aqueous solutions) were encountered within the salt. Such a brine pocket would make a repository unusable. In addition, exploration in the area showed that the bedded salt deposits are locally cut by what are called breccia pipes. These cylindrical features cut through the salt layers and contain rock fragments from the various layers. These "pipes" show that water has locally dissolved the salt layers and has created potential channels for ground water. Therefore these channels are obviously not acceptable in a repository.

Such findings demonstrate that careful study can almost always be expected to reveal features that demand special treatment or that rule out some otherwise promising sites.

The search for a suitable site in bedded salt has moved to the Palo Duro Basin in West Texas and to the Paradox Basin in Utah. The geological reviews of these basins are directed at finding sufficiently thick salt beds located at depths greater than 500 metres below the surface. Where these exist, the search will be narrowed by selecting only those deposits in which the beds are nearly horizontal and in which breccia pipes, brine pockets, and other indicators of salt-bed dissolution are absent.

Sites that survive these tests and that also lie in areas of low-population density are likely to be viewed favourably. If it can also be shown that these sites are not likely to be good prospects for oil, gas, or other resource exploration, they may become prime candidates for designation during the time period 1987-1990 in which the first two sites are supposed to be named. Once sites reach this "candidate" status, they will be subjected to a careful performance analysis. The general features of such an analysis are described in the NRC isolation study. A sample of the NRC preliminary analyses will illustrate what will be attempted in the late 1980s and early 1990s, as discussed below.

The goal of performance analysis is to show that radiation dose might be received by humans ingesting waster from a surface-water or ground-water supply above the repository. Such an analysis assumes that ground water reaches the canisters. The analysis then proceeds through the following steps:

1. Calculate the rate at which different radionuclides will be removed from the canisters.
2. Determine the shortest possible ground-water flow path from the canister to the surface.
3. Calculate the ground-water travel time.
4. Calculate the delay in the movement of each radionuclide due to sorption.
5. Calculate the dose that will be received by an individual from ingesting water that contains each radionuclide.

This figure shows the relative radiation dose delivered to an individual by each of the nuclides indicated. This graph was prepared on the assumption that ground-water flow would occur at a rate of 1 m/yr and that the path length would be

1000 m. Thus the time required for ground water to travel from the canister to the target would be 1000 years. Some radionuclides would travel at almost the same rate as the ground water; others would travel much more slowly. This graph illustrates two important features of such systems:

1. The highest relative dose comes from carbon 14, which is also the first radionuclide to arrive. Carbon 14 is not sorbed en route, so its travel time is about the same as the travel time for ground water. It contributes to a high dose rate until about 10,000 years after it first begins to arrive; then its contribution drops to zero. This is explained by the ultimate decay.

2. Actinides like neptunium do not arrive until much later, because they are sorbed en route. Because neptunium has such a long half-life, however, it contributes almost as much to the radiation dose when it does arrive as carbon did 90,000 years earlier.

The graph illustrates that a repository is capable of delivering radionuclides to the surface in separate episodes over long periods of time. This is the basis for concern with the siting and design of a repository. This concern will no doubt find ample expression in careful research and in the environmental impact statements that will be made for each candidate site in the 1990s.

Changing Attitudes Toward Contamination—Laws and Practice

During the years immediately following the Second World War, many bodies of surface water were allowed to become very seriously containated. By the late 1960s and early 1970s, changing attitudes began to be reflected in legislation to deal with this contamination. The initial concerns were with surface water, but more recently concerns have been shown for groundwater and soil contamination. In what follows we draw upon the survey of water-quality laws prdvided by the Council of

Environmental Quality in its annual reports to show the changing concerns with the contamination of soil and water (Council on Environmental Quality, 1981).

Efforts to improve the quality of the surface waters of the United States began with passage of the Federal Water Pollution Control Act Amendments (FWPCAA) of 1972 (33 USC 1311b2B). This act specified that all municipal sewage treatment plants were to provide secondary treatment of sewage by 1977 and were to employ the best available technology (BAT) by 1983. This law directed the Administrator in the Environmental Protection Agency to prepare or develop comprehensive programmes for "preventing, reducing, or eliminating the pollution of the navigable waters and ground-waters (33 USC sec. 1252a)". The Council on Environmental Quality observed that most cities had not met this goal by 1981. Thus the achievements reported here for the Willamette River illustrate what was intended by the law.

The FWPCAA was amended in 1977 by passage of the Clean Water Act, which was introduced to demonstrate the growing recognition that toxic contaminants from industrial sources were not being adequately controlled by the 1972 law. The Clean Water Act introduced a National Pollutant Discharge Elimination System requiring that each contributor to point-source discharge obtain a permit from the EPA and by so doing inform the EPA of the magnitude of character of discharge from each point source in the country.

Then it was recognized that even the complete control of point-source discharge from sewage treatment plants and industrial users might not be sufficient to achieve the desired end. This recognition was based on growing understanding of the importance of nonpoint-source contamination, which was not at all regulated and which was perhaps not easily controllable in any case. The EPA has no explicit authority to regulate nonpoint-source discharge, so it has chosen to operate under Section 208 of the Clean Water Act to provide for research on the contributions of urban storm runoff and agricultural runoff and to provide aid to municipalities for establishing plants for controlling such runoff. In addition, the U.S. Department of Agriculture is operating model programmes to show what might be accomplished by controlling agricultural runoff.

Although there is a long record of legislative efforts to deal with surface-water contamination, comparable efforts to deal with ground-water contamination have only begun. The EPA chose to concentrate on the then more immediate and readily demonstrable problems with surface water.

The problems of ground-water contamination were recognized with the Safe Water Drinking Act of 1974 (42 USC sec. 300f to 300j). This law requires that the EPA set drinking-water standards. The only actual provisions for exercising control over ground-water are contained in the Underground Injection Contral Programme and in what is called Gonzalez Amendment.

Research efforts have begun to deal with the Underground Injection Control Programme and the Gonzalez Amendment. These are directed toward identifying major aquifers and developing hydrologic data and models for each. Given these findings, the EPA will be able to determine where underground injection of wastes must be sharply limited. It will also be able to determine if a particular aquifer is what the Gonzalez Amendment calls a sole-source aquifer. Such an aquifer is one on which existing water users are totally dependent. The EPA has limited authority to prevent actions that might contaminate such an aquifer.

As contamination from hazardous disposal sites has become better understood and more widely recognized, legislation has been introduced to deal with this also. This legislation includes the Resource Conservation and Recovery Act of 1976 and the Toxic Substances Control Act of 1976, commonly referred to, respectively, as RCRA and TSCA. Both of these acts address the regulation of point-source discharge of toxic substances.

They are not concerned with contamination that has been identified.

The one other major federal law that deals with ground-water contamination is the Surface Mining Control and Reclamation Act of 1977. This act recognizes that surface mining can have disruptive effects, both chemical and physical, on ground-water systems, and it spells out the efforts to be made to mitigate these effects.